Kick-Starting Government Action against Climate Change

With drastic action needing to be taken now, rather than over the 30 years to 2050, this book addresses the crucial question of how to get action from governments who will always put short-term considerations (e.g. post Covid economic growth) over longer term climate priorities – unless forced to do otherwise.

How might governments be persuaded to implement policies that will result in effective action? And how can this be achieved at an international, as well as national, level? These are the questions that this book focuses on. Taking a systematic political science point of view and drawing on collective choice and other theories of political action, this book analyses the key political and economic dynamics shaping climate policies around the world, identifying major political opportunities that can be exploited by well-informed and determined political actors, such as NGOs and social movements.

This book describes how to advance and accelerate climate action around the world and will be of interest internationally to climate change campaigners, activists, political and environmental scientists.

Ian Budge is a political scientist who has pioneered the use of quantitative methods to study party democracy across countries. Currently Emeritus Professor of the Department of Government, University of Essex, he has been Professor at the European University Institute, Florence, and visiting professor at various institutions in five other countries.

Routledge Advances in Climate Change Research

Teaching Climate Change in the United States
Edited by Joseph Henderson and Andrea Drewes

Climate Change Law in China in Global Context
Edited by Xiangbai He, Hao Zhang, and Alexander Zahar

The Ethos of the Climate Event
Ethical Transformations and Political Subjectivities
Kellan Anfinson

Perceptions of Climate Change from North India
An Ethnographic Account
Aase J. Kvanneid

Climate Change in the Global Workplace
Labour, Adaptation and Resistance
Edited by Nithya Natarajan and Laurie Parsons

Governing Climate Change in Southeast Asia
Critical Perspectives
Edited by Jens Marquardt, Laurence L. Delina and Mattijs Smits

Kick-Starting Government Action against Climate Change
Effective Political Strategies
Ian Budge

The Social Aspects of Environmental and Climate Change
Institutional Dynamics Beyond a Linear Model
E. C. H. Keskitalo

For more information about this series, please visit: www.routledge.com/ Routledge-Advances-in-Climate-Change-Research/book-series/ RACCR

Kick-Starting Government Action against Climate Change

Effective Political Strategies

Ian Budge

Routledge
Taylor & Francis Group

LONDON AND NEW YORK

from Routledge

First published 2022
by Routledge
2 Park Square, Milton Park, Abingdon, Oxon OX14 4RN

and by Routledge
605 Third Avenue, New York, NY 10158

Routledge is an imprint of the Taylor & Francis Group, an informa business

British Library Cataloguing-in-Publication Data
A catalogue record for this book is available from the British Library

Library of Congress Cataloging-in-Publication Data
A catalog record has been requested for this book

ISBN: 978-1-032-11812-3 (hbk)
ISBN: 978-1-032-11814-7 (pbk)
ISBN: 978-1-003-22163-0 (ebk)

DOI: 10.4324/9781003221630

Typeset in Times New Roman
by Newgen Publishing UK

This book is dedicated to climate campaigners at all levels in the hope that it will provide more ideas for pushing governments into effective action against climate change.

Contents

Preface
Present dangers, present actions

The Covid-19 pandemic of 2020–21 shows how quickly our present socioeconomic and political arrangements trigger natural catastrophes, threatening our very existence on this earth. The more global these arrangements, the more the threat from unlimited trade and travel grows. This particular virus may have been contained – just. But who knows what the next will bring? And what other self-inflicted threats loom over us?

There is a close parallel here with climate catastrophe. Our global industries and economies are triggering a reversal of the natural processes that have, up to now, made planetary living conditions tolerable. Once these processes reverse themselves, we will be unable to reverse them back. Our current socioeconomic and political arrangements are enough to set the world heating up. But we cannot then stop it. Only if we take political action now to reform our collective behaviour can we survive in anything like the shape we are in.

REPENT FOR THE END IS NIGH! Taken out of a Biblical context, that is the message of this book. It is, however, directed more at governments than individuals. As the Covid crisis shows, governments are the key policy-making bodies that can initiate action on behalf of all of us. What individuals can and must do is make governments act immediately to stop temperatures rising to unprecedented levels in the 2020s – a process which has already started. Forests from the tropics to the sub-Arctic are already burning on a vast scale, emitting carbon rather than absorbing it. With the increased heat more forests burn, starting with the most vulnerable continent, Australia, in 2019 and spreading every year through California. Sub-Arctic fires are also spreading, to peat bogs as well as to forests, contributing to a meltdown of the Polar ice caps themselves. Instead of reflecting the sun's heat back into space, the newly exposed surfaces absorb it. Meanwhile the oceans covering four fifths of the globe heat and flood, a process that will end with frozen gas deposits from their depths burning on their surface. All of these effects, set off by human emissions, are accelerating in a vicious cycle beyond our control. The consequences in the shape of unpredictable weather, floods, famines, animal and plant extinctions, new pandemics (and much, much, more!) add up to an apocalypse of Biblical proportions.

Having first denied the evidence for climate change, most politicians and businessmen have anchored themselves in a comfortable interpretation of what is going on. That is, that natural processes will remain stable in the face of human emissions, still absorbing most of the heat-transmitting gases, while we slowly bring these under control by reducing our own carbon output to zero by 2050. This comforting long-term perspective absolves governments from any major effort now. Aspirations for reducing emissions can be expressed and climate emergencies declared, while party governments get on with the really serious business of restoring national economic growth after Covid, through increased emissions if necessary (Chapters 2 and 3 below). The problems of taking environmental initiatives are compounded by having to get the agreement of around 200 country governments for any concerted global action.

Scholarly and scientific attempts at puncturing government complacency have mostly consisted in making an ever-better founded case for immediate action, but leaving it at that, as though the accuracy of climate science itself is the problem rather than the political will to act on it. The few analyses that have focused on climate politics (listed in the end Bibliography) have mostly viewed them in abstract terms – getting agreement on negotiating procedures rather than kick-starting concrete action here and now. Drawing on professional research on party and government behaviour, this book discounts the possibility of getting unforced action to reduce carbon emissions. Instead, it focuses on how governments – even non-democratic ones – may be pushed into taking scientifically-based counter action by the popular and expert pressures specified below.

The likely switch from climate-focused action to promoting economic growth after Covid is all the more tragic since the necessary uplift could be provided by tackling the wider environmental errors contributing to climate change. To name only a few: leaking water supply systems and polluting sewage releases into rivers need to be fixed; landfill sites can be mined for now scarce materials and their methane emissions channelled into clean energy production rather than released into earth and atmosphere; carbon-absorbing reforestation and rewilding of the countryside can provide jobs at all levels, often in poorer regions where they are most needed; house insulation can be undertaken everywhere, since outside Scandinavia something like 40 percent of harmful carbon emissions are domestic.

All these positive initiatives would, by creating jobs, reduce demands on another climate and economy-supporting measure – the minimum income guarantee discussed below. Both this and environmental reconstruction are better ways of providing an economic boost than military expenditure or giving away public money to banks and businesses to hoard or invest as they think fit ('quantitative easing').

An environmental 'New Deal' runs up against the ideological shibboleths of public finance – 'public bad, private good' and 'balancing the books' – considered in Chapter 3. These beliefs, however, have been suspended – even by governments endorsing them – during the Covid pandemic. They also run up against the latest economic thinking embodied in new monetary theory, that responsible governments can build up financial confidence with necessary projects. Indeed, lenders in 2021 are currently paying governments to take their money, with zero interest prevailing during the Covid emergency. The new international agreements of June 2021 on taxing multinational corporations realistically provide additional financial support for environmental reconstruction. International action, however, is undermined by the ability of each government to plead its own special case for carbon-increasing developments, while blaming others for their consequences (like Australia opening up new coal deposits in the aftermath of the 2019 fires that destroyed an eighth of its forest cover).

Actions by individuals to reduce warming are limited and ineffective if governments are not prepared to support them and act against those who will not cooperate. Even the technological advances that have developed in part response to the climate crisis and can reduce our carbon emissions – solar, wind and hydroelectric generation, electric cars, house insulation – are not going to be applied quickly enough unless governments enforce their use and ban carbon usage. This is as true of targets set for 2050 as it is for the crisis already with us in the 2020s.

Even relaxed long-term targets are not being met – or even set – by more than a third of world governments, in spite of international climate conferences being held with increasing frequency. Holding these (in spite of some real achievements) is part of a general political smokescreen of promising climate action while in practice delaying it. But even the crucial World Climate Conference (COP26) due to be held in 2020 was postponed for a year by the British Government, while US adherence is hugely dependent on which party holds power in Washington, with Republicans denying climate change altogether.

In the face of such political inertia what can individuals do? It is tempting to conclude that nothing *can* be done – just enjoy the economic benefits while we have them. Most folk, however, having just experienced one world crisis, want to do what they can to avert the next – if not for themselves then for their children. Their problem is not knowing what individual actions will make a difference. The current alternatives on offer – voting Green in elections, mass demonstrations, direct action against oil/gas extraction – have not stopped climate-changing emissions over the last 30 years nor substantially protected the environment.

However, there are other ways in which individuals – acting together as well as individually – can make a real difference. A first essential is to organise action on a permanent basis so that potential campaigners have continuing guidance and help on what to do about climate disaster. A good precedent lies in the social crusades of the 19th century, particularly the organisations that committed their adherents to a written pledge to take specific steps within their own control (e.g. in the case of the Temperance Movement abstaining from alcohol). There is no reason why national organisations should not sign up individuals to a list of similar pledges – to cut their carbon and plastic consumption; eat less meat; maintain wildlife gardens; abstain from cruises and air flights; insulate houses and so on. Individual resolve could be strengthened by websites and individual tweets, local branch activities and meetings – all to reinforce collective action and solidarity. Individual subscriptions and crowdfunding could also be used to finance the national and international initiatives detailed below.

Another precedent for collective action is more directly political – the mass boycotts of apartheid South Africa in the 1980s. Boycotts could be organised effectively against countries burning rainforests by pledging always to enquire (and refuse to buy) products containing palm oil or South American beef and to push supermarkets and distributors selling them into action against them, at board level as well as middle management. Another action would be individuals pledging not to take plastics or products contained in plastics among their purchases, and/or to dump them on the supplier if they do. Investors could also check whether companies they have shares in are ecologically responsible, and either withdraw their money or agitate for change at annual general meetings and/or in written communications. (Successful action of this kind elected two activist Directors to the Board of Exxon, a notoriously polluting US oil corporation, in May 2021.)

Many other types of initiatives by individuals and companies could be undertaken. With mass support these could fill in a bit for governmental action, or, like their 19th century predecessors, actually pressure governments to act. The first essential, however, is to organise so as to harness individual efforts on a continuing basis to make a collective and sustained rather than a dispersed and sporadic impact on the changes going on. This is not to say that existing ecological movements have not been highly successful in securing media coverage and especially in influencing public opinion, But they are constrained in actually implementing their policies by the political institutions and procedures they work under. These are mostly designed to hamper rather than encourage new policy initiatives.

One limitation on all climate-related campaigning is its appeal mostly to non-manual, educated professionals who have both the time and energy to take far-reaching political action. This is a limitation that must be overcome by reaching out in an organised way to the general public (Chapter 5). It is also true, however, that middle class groupings provide most of the tax basis that governments rely on to finance their activities. This gives climate change opponents a substantial if not decisive clout in dealing with them by withholding payments in perfectly legal ways for activities of which they disapprove – but only if they organise to do so. Chapter 6 suggests how to do this, while still keeping such action within democratic bounds and not hurting public services.

However, it is always necessary to get the larger public onside. The problem is how, given that many live under socioeconomic circumstances that make feeding their family the overwhelming priority, rather than anything that might affect them next year. To give ordinary people the security to take a long-term view, Green parties and ecological movements need to counter the economics that insists that 's/he who does not work (or loses their job) shall not eat'. They can reach out at both the ideological level (Chapter 3) and practically, by campaigning for a permanent minimum income guarantee for all. (Incidentally ensuring a 'just transition' for those who lose jobs in an ecological clean-up.) This would extend the measures already adopted in many countries during the 2020 pandemic and shown to be feasible then. What a long-term income guarantee might look like is discussed in Chapter 5.

Divorcing life chances from jobs destroys the main justification supporting carbon emissions. This is that environmental degradation is necessary because national economic expansion and growth are the only way to finance any social or environmental improvement. A massive environmental regeneration – an ecological New Deal – and a decent income guarantee for losers would be alternative paths both to economic recovery and clean growth.

Income redistribution and cutting excessive consumption, rather than climate-damaging economic growth, are the real necessities for recovery after Covid. This would give climate change campaigners, particularly Green political parties, a winning electoral message for those who cannot afford to take climate change into account as things stand. These points are expanded in Chapter 5. Again, political organisation is required for popular mobilisation behind these issues.

All this demonstrates that organised individual actions – organised, coordinated and with clear plans for immediate collective initiatives – can have a great impact. It is still true, however, that national governments are the key, since they ultimately control developments within the territorial units into which our world is divided. In the absence of a world government to promote planetary action, national governments can also influence and coerce others into taking climate calming initiatives (e.g. governmental boycotts of goods from environment-destroying countries can powerfully supplement the individual ones discussed above).

However, given their current willingness to abet climate-changing developments and unwillingness to regulate them – all in the name of cheap imports and fostering growth – governments must be pushed by their citizens into action. This makes democracies and their governments the key political arena for defusing heat emissions. Their extended freedoms render citizen initiatives possible, even if we have to push the political boundaries somewhat to do this (Chapter 6). Democratic governments can also be pressured to act internationally, again prompted by citizen initiatives.

The following chapters go into detail about this, starting with the universal scientific consensus that absurdly little is actually being done to tackle climate change, even on the most optimistic incrementalist view. The catastrophic reversals happening now render action even more imperative – a matter of basic survival not only for humans but for everything in the world as we know it. This book details the collective strategies we need to avert this, starting from the fact that democracies are central to effective world action, and that political parties and governments are the key actors within them.

Further reading and reflection

Limitations on individual action to cut emissions are illustrated by reports in the *Independent* newspaper of 14 March 2020, pp 28–29, where four professional workers reported their experiences in trying to live on a carbon emissions budget of one tonne annually (averaging 2.74 kilogrammes per day). This is what is calculated to be the average sustainable level of emissions per person in the world by 2050. The major items where cuts counted were housing, food and travel. All the people involved lowered their house heating and went on a vegetarian diet. While they walked or cycled as much as possible, their jobs mostly required public transport to get to work, or air travel to visit business contacts or to attend conferences. Such travel blew great holes in their carbon budget, while lowered heating was distinctly uncomfortable in most cases owing to a lack of proper house insulation.

This illustrates that even laudable action by individuals is bound to have limited effects in the absence of direct government interventions (ending transport subsidies, particularly for air fuel, legislating for more home working to avoid travel; environmental restoration; raising building standards and subsidising insulation for older houses, to name but a few). Individuals should do what they can. But major results can only be achieved by banding together in sustained collective action, particularly political initiatives, to force governments into reducing emissions both domestically and internationally. Without continuing organisation and pressures from the grass roots they will always prioritise other short-term concerns and ideological responses (see especially Chapter 3 on this). A shocking statistic is that the unsustainable decline in polluting activities during the Covid-related lockdowns of 2020 only reduced harmful human emissions by 17 percent worldwide when the goal for the decade has to be 100 percent!

A masterly overview of the interlinked climatic and environmental threats is David Attenborough's *A Life on Our Planet* (London: Witness Books, 2020). This emphasises comprehensive restoration of the planetary environment as the only way to solve the climate crisis. As with almost all writing in this area, however, it details the science and specifies what needs to be done but not *how* to get it done. This book bridges the gap with its analysis of the collective political strategies needed to kick-start real action.

Acknowledgements

This book owes an enormous amount to the advice, support and encouragement of Jules Pretty, Professor of Environment and Society at the University of Essex, life-long environmentalist and climate campaigner, and to Professor Sarah Birch at King's College London on the social and political impacts of climate change around the world. Most strengths the book has derive from them and from my excellent Earthscan Editor Caroline Church.

I myself must take responsibility for failings or omissions but would be glad to have them pointed out to me at budgi@essex.ac.uk.

1 Forests burn, ice melts, seas surge, weather worsens

What governments have done and not done

1.1 World turning upside down

Since they improved steam engines at the beginning of the 19th century, humans have changed the world around them in increasingly significant ways. Small scale agriculture producing local food crops could certainly destroy forests and spread deserts as it exploited unsuitable habitats and moved on from exhausted soil. Commercial agriculture linked to distant markets by railways and steamers compounded such destruction, particularly as mechanised communications opened new areas and continents to exploitation. Vast settlements round mines and steelworks replaced open country with dense urban housing everywhere in the world, spreading pollution and gobbling up natural resources for which rural regions were cannibalised.

Such developments improved human wellbeing in practically every way – including the scientific knowledge that enables us to understand and tackle global changes. This combination of positive and negative effects from industrial development shows that we cannot advance into the future by going back to underdevelopment, as some climate activists want. Rather we need to manage developments so as to benefit both humans *and* the natural world. Science gives us the means to do this if we apply effective social, economic and political techniques – which is what this book considers.

In the 20th century developments of both a destructive and benign nature were slowed by crises of various kinds – economic recessions, military stand-offs and wars. They thus continued to have mainly regional and local effects, leaving oceans, polar regions and major rainforests untouched, lacking the technology capable of penetrating them. Concerns about the environmental consequences of human activity were thus, up to the last quarter of the 20th century, largely expressed in terms of local effects on animal and plant species – hunting

DOI: 10.4324/9781003221630-1

to extinction, destruction of habitat and other serious but limited changes. This spurred movements in the developed world for preservation of beauty spots and natural landscapes, or for tackling localised pollution, particularly in big cities. Commendable efforts to control these were made through the creation of national parks, conservation areas, planning regulations, trusts and voluntary bodies, national and local legislation and other protective strategies such as community buy-outs and empowerment.

Nobody could deny the cumulative and continuing benefits of such measures for the world environment. To some extent, however, localised successes helped mask the more ominous developments going on at the planetary level, in terms of the gases that surround and sustain us all (but could also kill us). The oxygen that makes up a fifth of the atmosphere surrounding the Earth is the basis of all life on it. Where it is replaced by carbon dioxide, as on Earth's neighbour Venus, the heat it stores from the Sun, and conserves on the surface with its clouds, renders life impossible. Human activity now pumps carbon dioxide into our air, along with other heat conducting gases such as methane. Amounts have increased exponentially since the Second World War ended in 1945, with extended use of oil and coal for electricity generation and other industrial uses, house heating and cooling, and resulting gas emissions. The near universal use of petrol burning vehicles (cars, aeroplanes, tractors) has also built up the amounts of heat-conducting gases in the atmosphere. Replacement of all these by processes that emit no carbon but take advantage of the natural processes already operating around us (wind and solar electricity, electric cars) is proceeding quickly – but not quickly enough to stop dangerously increased emissions of greenhouse gases.

This is true even if we take a relatively comfortable incrementalist view about current developments. Even if we see the goal of zero-carbon human emissions by 2050 as solving the problem, progress towards it at the present time is not sufficient. Many governments off-load their responsibilities on to others and leave technology-based reductions in emissions (e.g. electric cars) to industry.

The situation is much more alarming if we consider the current evidence on climatic developments. It points to a sudden reversal of natural processes, such as the absorption of carbon dioxide by forests, bogs and oceans: currently, gas absorption seems to be reversing into emission of the gigantic amounts of gases these have stored up. Such emissions have produced the extreme weather events of 2019–2021 – fire, droughts and flooding, with tremendous social consequences in the shape of pandemics, famines, mass migrations and wars. Current human activities push the natural processes of gas absorption and heat reflection to a tipping point where they turn malevolent rather than beneficent. Controlling human emissions, when it happens, will not be sufficient – either by neutralising or reversing them – to change natural processes back to where they were. Their tipping point is also the point of no return. If we go beyond it, we will live in a permanently changed world where mounting heat is the controlling physical reality.

Moreover, these irreversible changes will occur over a very short space of time – on present projections the next ten years (2020–2030), that is to say in our own lifetimes, not just our children's. We will all have to live – or die – with their devastating impact. The geological record shows fundamental climate reversals occurring within very brief periods. This is explained by natural processes suddenly reversing themselves as they are doing now. From this point of view, we have only five years to make a crucial difference and, literally, save the world. If this seems alarmist, even a more relaxed incrementalist view of what needs to be done in slower stages brings us to the same conclusion, since even what minimally needs doing does not currently get done.

Drastic action needs to be taken immediately. And it can only be taken effectively, as in the Covid emergency, by governments because that is what, ultimately, they are there to do – decide, enforce and organise the collective social action, which no other body is able to undertake. Some governments with autocratic or dictatorial powers are capable of deciding and enforcing such action on their own, once convinced by scientists of its necessity. China is a good case in point. It has already done much in the way of reducing its carbon footprint, though not in ways democrats would approve.

Democratic governments, on the other hand, have to respond to a public only half convinced of the necessity for action, hooked on the idea of gradual climate change on which action can be postponed indefinitely. Australia is the salient case here. Despite fires devastating forests over an eighth of the continent, with a three-month pall of smoke over major cities and 3 billion animals incinerated – and now unprecedented flooding – its government (recently re-elected over an environmentalist opposition party) continues to open up coal reserves for export, claiming that national prosperity and economic growth must go on unabated to make up for the economic losses caused by the climate disaster they produced. Many other governments (e.g. in Poland) also take such a stand, as they drive economic growth with the most polluting kind of brown coal.

Such governmental attitudes constitute the major obstacle to immediate climate action. Only governments can initiate and enforce effective action over their own territory, since they are the only political actors there with the ultimate authority to enforce new laws and regulations. Only they are able to push everyone into obeying them, through their powers of coercion. And only they are able to raise vast sums for necessary measures.

Normally in democracies harsh climate facts are masked by recurring elections and surrounding debate and argument that encompass change deniers as well as those urging action of various kinds. In this situation, governments find it easy to shift responsibility for inaction on to others, meanwhile getting on with their own pet projects. These involve 'growing the economy' in the old way, regardless of environmental costs. As Chapter 2 will show, however, they cannot, just as with wars or pandemics, wholly abdicate responsibility for organising the national effort in the face of approaching disasters, and getting other governments to do so too. From this perspective, we may ask what national governments have done to confront climate disaster? And, more pertinently, what can they do now to prevent it in the 2020s? These questions are considered in the next two sections before going on to the wider question of how to push governments into doing much, much more to confront the emergency.

1.2   What have governments done to confront climate change?

Governments, whether at country, regional or international level, are complex organisations with many actors involved. Nominally unified under the leadership of a cohesive political party, it turns out that the party itself is often split between various factions, each claiming to represent its true voice. Even where a cabinet or government coalition reach an internal agreement, it still has to push its policy through parliament, courts, groups representing various interests and warring ministries and departments, with a possibly inefficient or corrupt administration on the ground. Institutional struggles may be compounded by generic differences of opinion over what to do and how to do it. These difficulties indicate that we should never take governmental pledges and targets at face value or equate them with actually implemented policy, as over-optimistic climate analysts are prone to do. There will always be a gap between promises and real action, which needs to be taken into account in assessing actual progress.

Internal conflicts often result in the left hand of governments doing something different – or even contradictory – to the right hand. Thus, China's major contributions to dampening demands on the environment through limiting families to only one child, or by installing carbon-neutral solar panels to produce much of its electricity, have been accompanied by a single-minded pursuit of immediate economic growth by any means, including air pollution on a climate-changing scale and massive imports of raw material that have devastated forests from Siberia to New Guinea. Western governments have also faced two ways, for example by encouraging low emission rail travel while subsidising polluting road and air. Concentrating on the positive side, however, we should note that governments in general have:

1 Increasingly incorporated climate and environmental consider-ations into political and social decision-making. These often get overruled in pursuit of economic growth and development, but at least they are now considered rather than being totally ignored. This tendency has been reinforced at both national and inter-national levels (not to mention regional and local) by the Paris Climate Agreement of 2015. There, around 170 countries agreed to set targets for human carbon emissions and publish plans for achieving zero emissions by 2050. In spite of weaknesses both in the text and from a third of individual countries' failure as yet to even publish specific plans (and US withdrawal from the Agreement at one point), the agreement's binding legal requirements strengthen the hands of climate campaigners in opposing dam-aging developments at all political levels, covering both debate and practice (e.g. legal actions against polluting developments). In line with this, government bodies like the British Financial Conduct Authority have announced plans to make climate-related finan-cial disclosure mandatory for all public companies. Instead of just ignoring these risks and environmental liabilities, therefore, their vast and increasing scale will have to be revealed in accounts and assessed in investments (e.g. in building houses on flood plains).

2 Governments have also subsidised some of the technological developments that are reducing current emissions. Heating houses is a major source, up to 40 percent in some countries. At various times and to a varying extent, efficient insulation has been subsidised. Research into non-carbon-emitting means of electricity generation – solar, wind, waves – has also been directly supported or subsidised by governments, and planning and land laws altered to accommodate their development. Carbon capture and storage in old oil wells and mines is also being encouraged. The technology to take out and store carbon from factory emissions is already there. To do so directly from the air (DAC – direct air capture) is much more expensive, however, and requires massive government

support, so it will come too late to prevent natural reversals in this decade.

3 Paradoxically the major contribution of governments to technological advances that reduce carbon emissions has come from military or military-related research, where cost is no obstacle and normal economic considerations are side-tracked. Solar energy generation – taking energy directly from the sun – was at the heart of the US space programme. Computers famously came out of military code-breaking in the Second World War, while the internet was a by-product of inter-governmental nuclear research. If atomic fusion rather than fission can ever be developed as a safer, cost-effective, carbon minimising technique of energy production, that will also be due to governmental research support.

4 Much that governments can do is not through direct action, but by general regulation and law making. The best example is probably Clean Air Acts, which have had a tremendous effect in controlling emissions while improving health. Charges on polluters (e.g. congestion charges on cars entering city centres) have also produced major improvements to health as well as reducing carbon emissions in some cases. On the other hand, this is an area where governments have often failed to act effectively (e.g. tolerating rigged emissions tests on cars where offending companies have barely been fined). However, regulation of other bodies as well as of its own agents is probably the most effective means governments have of enforcing policy. The neo-liberal dogma of 'private good, public bad', prioritising freedom for corporate enterprises rather than their regulation, has to be challenged if carbon emissions are to be controlled (see Chapter 3).

5 Taking advantage of the massive handouts given to industry and other businesses to keep them going during and after the Corona pandemic, some governments, notably the French, have made their grants conditional on reducing emissions. Given the support businesses will continue to require over the next decade, this seems an effective means whereby governments can indirectly intervene to meet emission control targets. Governments are probably also wise to stand aside from the actual development of non-carbon emitting electric cars for example, where companies have already decided they are the vehicle of the future and are actively developing them (but still dependent on public charging points for car batteries, which governments may or may not be providing).

6 Similarly, population growth generally goes down with higher income and development, and the concurrent empowerment of

women. On current trends, it can thus be left to take care of itself – possibly a realisation that led to China ceasing to enforce its one child policy after 2010. A problem here, considered in the next section of the chapter, is that population increases have a different impact in high consumption countries like the US compared to poor countries in the developing world. There, individuals consume 20 or 30 times less per capita, thus producing much less in the way of harmful carbon emissions. On the other hand, more children provide more security in developing societies as there is a greater chance of some surviving to provide for their parents in old age. Coupled with better health care this has produced population explosions in poorer countries like those of Latin America, which lead on to the destruction of rainforests for more subsistence farming. The solution here has been taken by governments as fostering national economic development to provide better for the growing population. Unfortunately, this usually leads to greater environmental destruction, greater inequality and increased population growth, compounding initial problems of poverty and overpopulation. It would be better to provide direct income support. This has the effect of slowing down population increases by providing security outside the family and again by empowering women if they are paid it directly.

There are finite limits to economic growth and development in developing nations, with not enough planetary resources to bring all populations up to the developed countries' standard of living. The solution lies rather in transferring superfluous wealth from rich countries to poor countries through direct payments to individuals – a minimum income guarantee. This would have the double benefits noted above. First, by raising standards of living and empowering women, it reduces population growth and consequent carbon emissions. Second, direct payments by the donors' cashiers to individuals would avoid the local corruption and diversions of funding that have caused foreign aid to be defined as 'the transfer of wealth from the poor of rich countries to the rich of poor countries'. This was an apt description of foreign aid at a time when it was concentrated on major construction projects such as dams and steelworks – meant to provide jobs but in practice further enriching the rich, who siphoned off the funds. Direct individual money payments administered by their own personnel are more usually made by charities or the United Nations to refugees in desperate situations where basic supplies are distributed through international agencies working on the ground. While regarded as emergency measures, such operations do provide a model for making income payments directly to individuals throughout the world on a regular basis, thus reversing incentives for the poor of developing nations to destroy their environment in order to survive, and for the rich to divert the money into their own pockets.

Universal improved education is of course a solution to many of these problems, fostering professional standards among the better off and better understanding of their situation among the poor. This is an area where aid has been reasonably effective, even though so much more could be done.

1.3 What governments need to do but have not done

1. Guarantees of a minimum living income are of course sadly lacking not only on the international scene but also within states. Only Finland has cautiously experimented with paying a random sample of citizens a monthly income to see what the effects are. And, of course, individuals as well as businesses had their incomes made up in many countries during the economic stoppage produced by Covid. That was a tremendous step forward for governments to take. The United Nations and the current Pope have proposed that this should be done internationally. These are, however, only temporary measures. Made permanent they would take up a very large share of national resources if extended to all citizens – a particular weakness for governments in the developing world.

However, it is not necessary to top up income for citizens who already have adequate support in terms of their job, business, pension or inheritance. So, what we should be thinking of is simply a government *guarantee* that if individual incomes fall below the national level for decent living they will be made up to it. This procedure is referred to as a negative income tax – instead of taking away income below the minimum level, the state levels individual income up to it (while still taxing incomes above that level). A universal welfare benefit is still provided since even richer individuals paying taxes have an insurance against total penury in case of unemployment or business failure. Of course, volunteers or home carers also benefit in circumstances where they are not in paid jobs (though socially useful ones).

Though discussed in earlier periods, the idea of a universal minimum income has been, in practice, excluded from practical consideration under the dominance of free market economics and neo-liberalism since the 1980s. Social benefits are grudgingly paid out in the developed world with much scrutiny and policing of those receiving them, on the general principle that those receiving state aid must always be kept poorer than those in a job.

Here we come back to the starting point of the current economic argument – that carbon emissions and environmental destruction have to be tolerated or even encouraged to provide the jobs people need in order to survive, since only these will provide a reasonable income for all. Hence the argument for state subsidies to build factories or fund airports to provide jobs, ignoring the fact that it is cheaper simply to pay citizens money directly rather than prop up damaging developments so that these can provide income.

All this stems from the neo-liberal philosophy, dominant in current political and economic thinking, that paid jobs – even those of mind-numbing drudgery – are better than pursuing your dreams on state payments, even if many businesses are based on these too. One example is 'quantitative easing', a technical term that masks vast purchases of near worthless bank assets by state financial bodies (using taxpayers' money) in order to give private banks capital to invest and thus stimulate economic activity and new job creation (always provided the private banks decide to do this with the subsidy). Quantitatively easing the incomes of the poorest directly – who can be guaranteed to immediately ease their lot by spending on food and goods – has only recently been considered as a temporary strategy during the Covid pandemic, even though it is a more efficient and less costly way of stimulating economies in general.

2. A compromise between paying out money directly to those in need through a minimum income guarantee and creating jobs, work and income for those able to take them up, is for governments to initiate an environmental 'New Deal' on the lines of the American work programmes during the depression of the 1930s. These aimed at providing socially useful works that would never have been started without government support, e.g. the electrification of the Southern States with the Tennessee Valley projects.

In Europe, the major economic stimulus was provided by accelerated rearmament towards the end of the 1930s, as fears of a second world conflict grew. Arms manufacture is a key example of the paradoxes and self-contradictions at the heart of neo-liberal economic reasoning. Its products – shells and bombs – are destined for quick obsolescence as they get blown away, destroying infrastructure that will itself need to be replaced, thus stimulating even more economic growth in the future. Meanwhile, current products autodestruct creating an even more immediate demand for replacements that will themselves need to be replaced as they blow up. From this perspective they are the ideal market product – limitless demand creating ever expanding supply.

Environmental reconstruction shares with armaments a potential for generating much economic activity with limited potential for immediate payoffs. Shells and bombs evaporate into thin air. Rewilding and other environmental projects that absorb carbon similarly change the landscape in ways that have no immediate payoffs, though with a potential for immense economic growth and social benefits in the future.

Whether positive or negative in terms of general impact, there can be no doubt that both armaments and the environment provide an immediate stimulus that spills over into the wider economy as a spur to further activity and a major and immediate source of jobs, income and tax revenues. Unfortunately, governments are much more likely to re-arm than rewild as a response to threatening planetary developments.

There is indeed little evidence of governments currently seeing or seizing the economic opportunities of environmental reconstruction. They may, however, be forced into such work creation simply to avoid social unrest as economies shrink after Covid. This offers both an opportunity and a challenge for environmentalists to direct government initiatives towards positive ends, rather than to those aimed at killing people.

3. These examples underscore the point that government intervention is most effective and wide reaching – for good or bad – when carried through in the shape of general policies or regulation rather than specific actions like helping an ailing steelworks to preserve local jobs or destroying ancient woodland to provide airports, roads, high-speed rail and so on. We have already considered one negative general policy in the shape of preferring any kind of private job creation to guaranteed minimum incomes, all in the name of boosting individual independence and freedom from state interference.

Unfortunately, this fine sentiment creates one of the major political problems of our time – the existence of an impoverished underclass ready to clutch at any political chance to improve their lot, or at least to get back at those they see as benefitting from the system and running its established institutions. Their desperation also pushes the climate-changing interventions noted above as peasants are forced to destroy their own environment in order to stay alive. In fact, a guaranteed income payment with easy access for those who need it would not only render governments more stable but also make their administrative task easier while cutting its costs. Creating a receptive rather than a hostile environment for social claimants renders detailed assessment of their claims, with a view to giving them no more than they are strictly entitled to, unnecessary. If claims are met on the basis of their income tax assessment, costly additional investigations are not needed. Administrative financing can be given instead to social care, where social workers can pursue their proper role of seeing that clients' money is spent wisely, rather than desperately seeking basic support for clients in a bureaucratic maze of separate benefits and entitlements, all designed to make payments more difficult.

A general income guarantee runs counter to the prevailing economic philosophy of free markets. But it has so many climate-healing as well as social benefits that the ideological battle needs to be fought and won if we are to avoid catastrophic change (see Chapters 3 and 4). Unless we win this argument, the impoverished masses will destroy the rain forests and oceans in search of subsistence, or support the agribusinesses providing jobs by doing so, while attempts at reversing such processes will founder if they are seen as hostile rather than beneficent. Of course, a universal minimum income guarantee needs to be administered directly by international agencies (otherwise it will be diverted and embezzled in many places) and needs to be supplemented by importing countries refusing goods not made under regulated conditions like those required from its own producers. More detail on all these points can be found below.

4. A minimum income guarantee has such benefits administratively, by replacing and simplifying so many specific payments, that it is surprising that governments do not recognise them. Of course, there will be opposition to simplifying a tangle of different (and often unequal) state pensions, benefits and subsidies if their recipients feel that simplifying and unifying reforms are just a cover for reducing social payments (as with universal credit in the UK and President Macron's pension reforms in France). Replacing them with a simplified and usually more generous system for anyone needing it is surely a way of ensuring their acceptance and a just transition to reform, bringing with it greater social harmony and an end to wars of necessity against an environment that supports us all. How all this works out practically can be shown by its effects in one area – agriculture – one of the most important in terms of fighting emissions, and where governmental action to ensure their reduction is most notably lacking.

Agriculture of various kinds – crop-producing, herding, ranching, game and fish management – in fact covers most of the land surface of the globe. In one shape or form it feeds us all. Since it is territorially based, it is organised primarily round privately owned farms, ranches and estates, from miniscule peasant holdings with insecure tenancies up to vast latifundia covering thousands of acres, often operating as agribusinesses processing food and drink as well as producing it, managing house-lettings and hotels and organising tourism. Developments in this direction have intensified over the post-war period and produced powerful political movements and interest groups purporting to protect small farms but actually dedicated to advancing big agri-business.

Indeed agriculture, like fishing, is strongly divided between peasants and small traditional farmers struggling to survive (and often the intended beneficiaries of Government aid, subsidies and regulated high prices) and large-scale cultivators who claim to speak for all farmers but who frequently force the small men out. Since subsidies generally reward higher outputs, any increase benefits the big holdings more than the small. However, it is convenient for the large landowners to represent government aid as benefitting farmers as a whole, as if there were no conflicts of interest among them. The European Union's Common Agricultural Policy is a prime product of this type of reasoning, rewarding food output however achieved, in spite of palliatives introduced in recent years to let land lie fallow to protect wildlife, and regional payments directed to poor areas where small farmers are in the majority. Despite these, smaller farms continue to decline since indiscriminate subsidies of any kind benefit the big holdings more.

This is partly because they pursue increased output above all else, whether by creating enormous prairie-like fields where machines can work more efficiently (denuding the countryside and destroying soil in the process) or by facilitating monocultures of cash crops on an enormous scale, or by intensive chemical spraying of fertiliser and pesticide – aptly described as 'a continuing war against nature', since insects (the 'pests') are at the basis of the food chain and act as the general pollinators of plants and soil improvers. The need for more land leads to deforestation, often to clear the trees for small holdings, but then allowing the big entrepreneurs to force peasants into the remaining forest, to clear the next strip for their eventual takeover.

The excuse offered by agribusiness for environmental degradation and destruction – reversing the absorption of carbon into emitting it – is that 'we have to feed the nation'. A variant of the argument in more developed countries is that food production is better regulated there than abroad, so turning land to other uses (e.g. into marshes and soft sea defences for flood control) is counterproductive. The country will then have to import more food from less regulated environments, thus encouraging more environmental degradation and harmful emissions in the world than if they just left domestic agribusiness to get on with it. Such justifications conveniently ignore the food that could be saved by better storage (40 percent of all food produced is lost somewhere in the distribution chain); problems of obesity and illness caused by overeating, producing a heavy burden on health services; and probable changes in food habits from meat eating (livestock produce more emissions) to more of a plant-based diet.

In the face of agricultural lobbies using pseudo-environmentalist arguments, and mobilising even the disadvantaged small farmers against perceived threats to their existence, governments have done very little to reduce the disproportionate amount of taxpayer money they pour into agriculture. A simple solution, as in other areas of social and industrial life, is to end all subsidies and artificially high prices and replace them with a minimal income guarantee. Paid directly to all farmers who need to take it up, without being linked to levels of food output, this would make farming viable for traditional small-scale farming, more respectful of the environment in developed countries. It would also reduce the incentives for agribusiness to expand production at all costs, saving governments money in the process. A direct income guarantee would ensure that no money would be paid to agribusiness as an inducement to expand. Their profits are above the level of a minimum income anyway, so the guarantee would not be called in for them. Thus, money would only be paid to those in the sector who really need it and used to support more traditional, non-destructive usage of the landscape. The same argument applies to fishing fleets. All government subsidies should be ended and individual incomes topped up for the small folk.

The switch to minimal income support could be supplemented by reforms of general regulatory measures. A prime need here is to change from a highly permissive regime with regard to chemical-based agriculture – allowing all pesticides and fertilisers that cannot be definitively proved to damage the environment to be employed – to a more restrictive criterion – allowing only those that can be definitively proved to have only beneficial or neutral side effects. Other measures would restrict the number of livestock, mainly cattle and sheep, carried on each acreage. Scotland is a case in point where the countryside has been denuded of trees by allowing sheep and deer numbers to be double what they should be, while tolerating bad practices such as heather burning and extermination of predators and other species to encourage game birds. Again, the argument is that this promotes tourism and local employment. A guarantee of a minimum decent income eliminates this objection, which has paralysed necessary action to restore the environment that, of itself, would provide more paid work. Again, the same arguments apply to fishing in inshore areas and to fish farming supported on meal made from small fish.

Regulatory action is a way in which governments can reduce emissions without spending money, and even contributes to saving it! Regulation of unnecessary advertising would discourage production of many non-essentials that consumers are currently urged to desire and buy. A statutory provision, taking advantage of the computer revolution in office work, for all employees to have the right to work at a distance for all or some of the week, would vastly reduce demands on road, rail and even air transport. Employees could also live at greater distances from work, reducing construction and consumption in big cities and taking advantage of housing in economically declining areas, which then supports their regeneration.

The vast expansion in home working as a means of limiting the Covid-19 infection will surely foster this change. However, a statutory right to work from home would reinforce the social and economic changes from which governments so far have held aloof. Action of this kind would undoubtedly be opposed by those who see any regulation as a loss of business freedom, necessary for economic progress and growth. Querying this position is a political and ideological battle that needs to be fought (see Chapter 3). In some areas, however, the fight against emissions requires governments to withdraw from interventions that have the effect of encouraging them. Here the necessary actions are simple and direct – just stop! We review these in Section 1.4.

1.4 Just stop!

The pursuit of economic growth as a sign of national virility and a prerequisite for any social or climate initiative is often, paradoxically, used by governments to justify climate-damaging actions. The prime examples here are subsidies supporting the extraction of carbon emitting fuels. These include coal mining and burning, especially the very noxious brown coal of Eastern Europe and China; fracking, the injection of water under high pressure into shale deposits underground, to force oil and gas out; and extraction of oil from sands in Western Canada, described by observers as the world's single greatest environmental catastrophe. (It takes one barrel of oil to produce two barrels for sale.) We can add to direct support of extraction the subsidies paid to users, particularly by keeping petrol taxes low or tax discounts on polluting fuels such as the diesel used by aircraft. Resulting increases in air traffic then justify the building or extensions of airports and access roads, resulting in further emissions.

The list could go on. It has been estimated that direct government subsidies for carbon emitting fuels amount to US$5 trillion worldwide.

In any serious fight against the emissions promoting climate change these could be stopped immediately. Of course, this would be strenuously opposed by beneficiaries. The arguments in this case are always that the emissions in question are a relatively small amount of the world total, disproportionately benefit the country involved, or help some disadvantaged group or region or business within it, and, therefore, should continue to be subsidised until other countries stop doing so.

Such arguments are a specific example of the difficulties involved in collective action of any kind and particularly in cases where there is no final authority such as a world government to promote and enforce the general interest. We consider these problems, the domain of social choice theory, in Chapter 2. This is a question that particularly affects action against climate change. Every government wants the other governments to take necessary action against it, but also to prosper by being the last to do so or even being exempted altogether, so as to maximise all of its own benefits without suffering the consequences.

We should note, however, that stopping bad policies is slightly easier than initiating good ones. So, with increasing alarm over catastrophic change, there may be better chances of 'just stopping' them. The Extinction Rebellion – occupying central London for a week in 2019 and directly confronting world leaders – has highlighted this form of action. The other popular initiatives we discuss in Chapters 5 and 6 may be particularly effective here. But before coming to them, we have to consider the general problems facing all kinds of collective environmental and climate action (Chapter 2).

1.5 Government (in)action in the Covid crisis: a precedent for the climate emergency?

The world emergency that started in late 2019, and dominated the following years, focused on stopping a hitherto unknown viral infection of the lungs (Covid-19). Its rapid spread from China to the whole world owed a lot to globalisation and its encouragement of rapid mass travel round the globe. It was not, therefore, a direct consequence of climate change but rather a parallel effect from the processes producing it.

The pandemic did, however, constitute a sobering reminder of what consequences other types of pandemic, spread by migrations of insect populations as colder regions heat up, could carry for the world. The most effective defence against Covid-19 was quarantine, limiting human to human transmission of the virus. Quarantine, however, meant that people stayed home, so shops, factories and transport closed down. Long distance communication meant that much business and indeed research and other activities could be carried on from home. Obviously many others could not, such as manufacturing, shopping and services. Much normal life and economic activity just had to stop, with compensation for loss of profits and earnings paid for by governments. All this has important consequences for the fight against climate change:

a   Drastic reductions in carbon-emitting traffic – mainly cars and aeroplanes, manufacturing, shopping and economic activity in general – led to the kind of emissions reduction that climate activists had been campaigning for. They may even have allowed countries to really meet the annual cuts they signed up for in the Paris Agreement in 2015. Of course, the reductions are only for the pandemic years. In the absence of further changes emissions will bounce back again to dangerous levels. But reductions have been shown to be feasible. The collapse of a number of businesses will probably reduce national and world emissions permanently. A sobering thought, however, is that even with economic shutdown across the world, carbon emissions decreased by only 17 percent.

b   Unprecedented government financing of measures to avert social and economic collapse during Covid shows that other drastic action is feasible, along with regulations and measures that would never have been envisaged before the crisis struck. Quarantines and curfews are not necessary, of course, to avert climate change, but they do demonstrate that with public support new types of political action are entirely possible, thus setting a precedent for wide reaching environmental measures. By far the most important of these would be a New Deal aimed at repairing economies through massive environmental restoration: total renovation of sewage and water systems; cleaning of rivers, seas and oceans; rewilding of countryside; soft flood defences; re-creation of woodland and bogs; public re-charging schemes for electric vehicles; landfill mining and restoration; marine conservation promoting sustainable fishing – the list is almost endless. Such initiatives could supplement – and actually be easier to effect – than technological strategies such as sucking carbon dioxide from the air and storing it (DAC – direct air capture).

Emission-reducing economic growth should therefore be the aim, providing new jobs and technologies on a massive scale and promoting an integrated and environmentally-friendly plan for tackling the economic crash that the approaching climate catastrophe will also bring on its tail. Governments would have unlimited financial credit to do this by reassuring bondholders and lenders that they have a credible plan to avert future economic disaster, enabling them to guarantee their investments and pay interest from increased tax revenues from newly employed workers.

c In general, during the Covid-19 crisis, state regulation and intervention have been vindicated and shown as necessary – above all to ensure the operation of free markets! State regulation and intervention have been shown by an OECD study to have only limited or no effect on GDP or national productivity (*Economist*, 3 Jan 2015 p 55). This undermines the neo-liberal arguments that have held sway for the last 40 years, that the way to progress and prosper is to let markets operate with as little government intervention as possible, cutting public services in order to reduce taxes and placing short-term individual gains over future social losses. Instead, public services such as health, social care, environmental protection and police have been shown necessary to even short-term survival (and continued economic functioning) as nature strikes back. We will see in Chapter 3 how all this strengthens the climate side in the argument with neo-liberalism. Environmentalism has got to win ideologically in order to guarantee permanent government action of the kind we have been discussing.

Perhaps it took a catastrophe to bring about these mould-breaking government interventions, just in the nick of time to confront the pandemic. Unfortunately, the nick of time will not do for the climate crisis as the tipping point of non-reversable change may already have been reached! Negative consequences for climate change also follow from the Covid crisis, of course. These are:

a The first priority of governments after the virus has been controlled will be to restore economic activity to pre-crisis levels at all costs, including environmental ones. The old argument that conventional economic growth is the foundation for everything else will be revised, one prerequisite being looser environmental regulation. The Modi government in India has already passed such legislation allowing compulsory seizure of land for approved developments; the UK is substantially loosening its regulations for developers; and the US has already done so. Minimum income support will be withdrawn and most public services again downgraded.

b These moves will be encouraged by general complacency about having surmounted the present crisis and in the confidence that any future ones can be similarly overcome by last minute emergency measures. Climate change will continue to be regarded as a long-term problem with measures to tackle it being postponed in favour of short-term imperatives like building more houses, faster, and to poorer standards.

c Of course economic activity could be drastically boosted and job creation vastly ramped up by programmes for restoring the environment, creating a massive economic boom. New monetary theory supported by actual financial practices during the Covid crisis indicates that governments enjoy limitless credit if they retain the confidence of their bondholders. New international agreements on taxing multinationals should provide a solid foundation for this. Decisive action along with the immediate economic upsurge it would produce would also do so. Unfortunately, this path to economic recovery is unlikely to be taken by most governments who instead prefer to give money unconditionally to banks and private businesses for private investment. Post-Covid monetary inflation is quite likely to trigger austerity measures to cut government expenditures, particularly on public services, international aid and environmental improvement.

All catastrophes and crises carry conflicting consequences, just as conflicting priorities and strategies jostle for the attention of governments. Covid-19 is certainly a wake-up call for some but equally a reason for others to postpone climate action in the face of the short-term emergency. Activists cannot therefore hope it will automatically carry the day for them. They still have to fight for adequate and timely measures against catastrophe to be taken immediately. The first imperative is to win the argument with resurgent neo-liberalism, which is the focus of Chapter 3.

1.6 Further reading and reflection

The arguments for continuing to carry on the war against nature associated with large-scale commercialised farming – possibly the major source of carbon emissions – are well summarised in the following article from a local newspaper in an area of intensive farming, where hedgerows have been ripped up to make large fields suitable for mechanised contractors to operate, and with intensive use of chemical spraying. While starting off with a reference to the consequences of climate change in terms of flooding, it then uses them as an argument against soft flood defences and for the 'we must feed the nation' argument for commercial food production. Headed 'Utopian ideas failing in reality', it goes on to some of the arguments against climate action summarised in Chapter 1, so it is useful to report some of the text here:

> ... the awful floods in some parts of the country ... have caused stress and devastation to ... many lives.
>
> Surely those in charge of future policy can now understand that introducing beavers and using farmland as a sponge is an idea dreamt up at a Mad Hatters' Tea Party...
>
> ... governments have also been encouraged to pay huge subsidies to encourage organic farming.
>
> Recent research ... shows that a shift to 100 per cent organic production across England and Wales would reduce yields by about 40 percent, sucking in food imports to make up the shortfall. This in turn would require more land use overseas leading to a net increase in overall greenhouse gas emissions, even though UK emissions could eventually fall.
>
> Another study ... concluded that intensive, high yielding agriculture may be the best way forward to meet the rising demand for food, while conserving biodiversity by using less land ... 'the least bad way forward'.
>
> (Peter Fairs in the *Essex County Standard*,
> 28 February 2020, p 24)

The 'feed the nation' argument used here is likely to be undermined by British withdrawal from the EU with its protective Common Agricultural Policy (CAP) for commercial farming – probably to be replaced by a Free Trade Agreement with the US, following on a current agreement with Australia, whose large-scale agricultural production will overwhelm the British and introduce genetically modified crops on to the market. Arguments and evidence for the threat these pose are set out in a Soil Association booklet *Mother Earth*, Vol 7 winter 2012 – www.soilassociation.org/motherearth.

On a broader front, the UK's Parliamentary Committee on Climate Change, (www.theccc.org.uk) in its report of 2019 adopting an incrementalist view of change, recommended a net zero target for emissions for 2050 (for Scotland, 2045), but noted that a major strengthening and extension of current policies were needed to achieve even this distant goal. Its follow-up assessment of June 2021 noted that less was being done currently than when its previous report was published.

The report makes very informative reading on what detailed changes need to be made. The emergency measures taken during the Covid lockdown make some seem more feasible (e.g. home working), but many others are likely to be passed over or even reversed in the cause of getting the economy back to where it was.

A recent hopeful development is the government plan for post-Brexit agriculture in England, which cuts subsidies for large commercial farmers and pays for environmental improvements such as restoring hedged fields and returning marginal land to marsh and woodland, forming 'carbon sinks' trapping carbon dioxide. Whether this policy would survive the government's other aspiration for a Free Trade Deal with the United States involving massive imports of cheap food raised by environmentally hostile methods remains to be seen. The pilot deal with Australia has been approved in the face of internal government opposition.

Two newspaper reports emphasise the urgency of action long overdue in regard to deforestation in the Amazon rainforest in Brazil (*Guardian* editorial, 6 June 2020) and setting up marine protection areas in world oceans and seas (*The Scotsman*, Monday, 23 March 2020, p 27). An encouraging business-based initiative was taken in early 2021 by a consortium of British supermarket chains led by Tesco, warning the Brazilian legislature of the negative consequences of approving clearance of a vast section of rainforest.

To strengthen inadequate protective measures against illegal over-fishing and seabed destruction in the Dogger Bank (shoals in the middle of the North Sea), the militant environmentalist group Greenpeace dropped large boulders to snag bottom scraping dredges (September 2020), against the protests of large trawler owners. Besides their direct effects, such increasing corporate and private initiatives pressure national governments to back them up with legislative and administrative action of the type discussed in Chapter 2.

2 Problems of collective action

Policy making and enforcement

2.1 State, government and policy

If we are to have world-wide action against climate change, we have to rely on territorially-based governments to carry it through, either singly or by cooperation among themselves. One of the great contributions of post-war climate studies is the realisation that changes in any part of the world environment will affect all the others. So, for their actions to be most effective, governments have to take them collectively by holding global conferences to set emission targets and register binding commitments to meet them nationally.

Everything would be easier if we had a world government capable of enforcing global action, rather than country governments nego-tiating about it. As no single world authority exists, we have to have negotiations between governments. But this does not mean that these cannot make a difference by acting within their own territory, not only by enforcing international agreements but also by promoting helpful initiatives of their own, which others might copy, e.g. by refusing to import environmentally damaging goods and food from abroad and prohibiting national investment in their production.

Territorial state governments are necessary because individuals and voluntary groups are not capable of providing an ordered communal life all on their own. Individuals might join together to run a market for example, which trades goods and services and brings together buyers and sellers, enabling everyone to get what they need quite efficiently. When it comes to ensuring quality and enforcing order, so that jealous competitors do not try to lower standards or come to blows, the stall-holders might jointly formulate rules to keep the peace and ensure fair competition. However, some individuals and groups might not join, so saving on the costs of enforcement, since they would benefit from any resulting order and peace anyway.

DOI: 10.4324/9781003221630-2

Such 'free-riding' stems from the fact that public order and peace are 'public goods' that, if provided at all, must be provided for everyone, even for those who refuse to contribute to them. Other examples closer to the climate debate are flood defences or stopping air pollution. You benefit from the effects even if you refuse to pay. Faced with widespread non-co-operation the proposals may never take off. To provide an obvious public benefit, the state usually steps in, forcing everyone to pay by jailing or fining them if they do not.

More indirectly it will impose general taxes on the whole population. Then the government will decide what public projects need to be pursued and paid for. Such decisions are generally termed 'policies', general decisions about what should be done on the state territory and how to do it. Of course, a policy of non-interference might also be adopted in some areas. Deciding not to do something is just as much a policy as deciding to do it. There are no areas that are inherently free from state intervention, though it may be wiser sometimes to leave them to run themselves.

In the past, environment and climate have largely been left like this, though governments have intervened occasionally to provide national parks or banned chemicals proved to be dangerous. As in other areas of life, however, new problems mean that greater regulation and more extensive interventions are now called for. These will only have patchy success if governments do not back them. It is this that renders governments so crucial in the fight against climate change. Only they can initiate and finance truly general action against it.

But governments have to be stimulated or pushed to provide these public benefits and there are always powerful political forces opposed to any action. Private landowners resent their own use of land being regulated or controlled and their workers fear loss of jobs and income, so the immediate reaction to, for example, rewilding for emissions reduction is likely to be hostile. Pursuing long-term objectives at the cost of immediate unpopularity is difficult for governments, especially democratic governments who may be facing re-election.

Incumbent governments steadily lose votes during their term of office due to such 'costs of governing', i.e. alienating some electors with proposals to provide or withdraw public goods. Those adversely affected by the proposals are always quicker to react against them than those benefitting are to support them. Immediate losses are always more obvious to affected groups than long-term benefits. Hostile reactions are reinforced by party ideologies that see all state interventions into the private sphere as necessarily bad, so they always carry a presumption against it, which reinforces self-centred opposition. Where a party with these views controls government, it is hard to initiate any environmental or climate action at all. In democracies, however, there are other ways of forcing governments to act, which we now review.

2.2 Parties, agitation and democracy

Democracy has been characterised as 'a competitive struggle for the people's vote', with the competitors being political parties. Parties are in fact distinguished from other organisations and groups by running candidates for election to political office. These might be an executive president who forms an administration to govern for a fixed term, or a parliament where democratic governments have to be able to win a vote of confidence in order to form in the first place, and then to continue in office. Parties might get a vote of confidence because one of them obtains a (near) majority of seats in the latest election, or because like-minded parties join in a coalition to get a majority in parliament and form a policy-making government. This lasts until it fails to win an important vote and/or falls apart because of quarrels among its party members. In such situations an alternative coalition government might be formed that can win parliamentary votes, or another election might be held to initiate a new process of government formation.

This is a very broad sketch of the way democracies work. It does, however, correctly emphasise the centrality of their two distinguishing institutions: elections, which bring the populace into governing and policy-making processes, and political parties, which provide voters with reasonably clear alternatives to choose between and which then organise governments on the basis of the votes they obtain.

In order for the general election itself to be a free and informed expression of electoral preferences, it needs to have supporting safeguards, which involve electors having certain rights. The most obvious are free discussions at meetings or in newspapers and media without any relevant group or individual being barred; free voting choice without sanctions or reprisals; and being able to organise political parties and supporting groups to represent opinions as they wish.

Parties indeed are so important to the functioning of modern democracies that these are best described as party democracies. We have just seen how parties enable clear choices to be made in elections by offering electors competing teams to form alternative governments. Such choices might be made on the basis of candidates' personal appeal or their perceived competence, but they also involve some consideration of the policies they endorse. Parties themselves are generally held together by a central ideology that stresses certain priorities for action and predisposes the party to act in particular ways to achieve them (through more or less regulation of social life, for example). So, in voting for a party, you are also choosing a particular set of policies even though that may not have been the main consideration individuals had in casting their vote. Of course, circumstances may blow a party government off course, as when the 2020 outbreak of Coronavirus forced governments of all kinds into more extensive regulation and social intervention than some would otherwise have undertaken. Or stand-offs between presidents and legislatures or party factions or coalition parties may produce compromises that are not exactly what any of the participants would have wanted, so the actual policy implemented may well differ from party targets as advocated in the election. Over time, however, party policy targets will have some effect, thus linking the election result, even if imperfectly, to government actions.

In such a party-dominated situation, the most obvious way to promote awareness and action on climate change is to form a political party focused on its prevention, with a definite programme of action to confront it. And this, by and large, is what climate activists have done. Green parties have been formed with an organisational base comparable to that of established parties, that is, a national membership paying subscriptions that give the party an independent financial base; a network of local branches that discuss policy and canvass and organise locally to support party candidates; and a disciplined group of elected representatives in the legislature that advocates and votes for party policies. Such a party organisation is common across all parties regardless of what ideology or policies it is employed to support.

However, getting involved in 'normal politics' in this way has not paid off particularly well for the Green parties emphasising environmental improvements that might slow down climate change. 'Normal politics' demands that they must get involved in short-term issues such as constitutional reform, health care or preventing the break-up of the EU, not to mention tackling crises such as the Covid pandemic.

Such issues are of course important in themselves and may carry some potential for putting the Greens in a more powerful position for tackling climate change in the future, but they do take up a lot of time, resources and energy, which are then diverted from directly tackling the climate crisis. Moreover, Green parties have not, on the whole, succeeded in forming or entering governments since they first emerged in the 1970s. Certainly they were partners in coalition governments in Germany and Belgium round about 2000, while in Scotland they have been important sustainers of pro-independence governments. In return these have blocked further nuclear development and fracking for oils and gas, extended free bus travel and brought the date for zero emissions closer. These are admirable political achievements on the ecological side, but they do not substantially speed the central developments needed to tackle climate change effectively, such as reversals in land management. They do a useful and essential job in publicising and promoting these, but have so far not won over governments or a popular majority to actually carry them through.

Political action at the party and electoral levels thus gets climate activists tangled up in short-term concerns that sometimes give them an influence over governments, but only on relatively peripheral matters. To make a major difference they have to get massive increases of votes and seats. One can sympathise with their failure to do so, involving much long-term lobbying, patient negotiation and messy compromises. Green parties have done their best, but the political processes involved are so slow and tangled that they tend to divert them from their original mission of producing long-term policy change on climate to an involvement with the short-term concerns of the established parties. One can only get long-term change by dominating governments, yet to get into governments you have to take up short-term concerns. You might slip in one or two objectives of your own, as Greens have done, but these will not measure up to the climate crisis that actually confronts us.

It is of course possible that, buoyed by some governing experience and by a wave of popular concern as the climate crisis worsens, Greens will eventually take over governments and launch appropriate climate measures. But this will be achieved too late. The irreversible changes in natural processes are already underway and normal party policy-making is simply too slow to produce the immediate, urgent action required.

Is there another democratic way to speed up effective climate control? There has been a recent attempt in the shape of Extinction Rebellion. Manifestations in a number of countries have gone beyond the normal bounds of mass assemblies and street processions to blockade big cities, publicising the cataclysms that await us and urging governments to initiate immediate and fundamental counter action. Extinction Rebellion has supplemented normal demonstrations with prolonged agitation at all levels, from school strikes to heckling world leaders. These have undoubtedly publicised the climate emergency and raised popular concern. However, they have not pushed national governments into actually doing what they have not done (as listed in Chapter 1), nor have they even forced them into ending harmful policies. Governments, after all, are well experienced in confronting and controlling mass demonstrations, even violent ones. They saw off the Gilets Jaunes rioting in France in

2018–2019 and the violent ecological protests at G7 meetings of leaders of the richest world countries round about 2000.

It seems therefore that neither of the main methods currently employed to promote climate emergency measures – parliamentary and street action – have really succeeded. The dominance of free market economics and neo-liberal prejudices against public regulation and intervention have so far defeated them.

Paradoxically, however, governments' lax control of businesses in the national and global free market has opened up another avenue that climate campaigners could exploit more effectively. This is to use the tax loopholes and tax avoidance mechanisms hitherto only open to businesses and the rich. As the Salvationist slogan went in the 19th century, why should the Devil have all the best tunes? Similarly, in the 21st century, legal tax avoidance should not be left only to businesses in order to swell their profits, but used to divert revenues to environmental causes and put pressure on governments to do so too.

Reducing the tax base on which governments depend to finance their policies (sometimes good but often environmentally harmful or irrelevant) would really hit them where it hurts and push them to restore their position by taking positive and effective actions against climate change and stopping their harmful ones. How to do this demands considerable reflection, mustering of support and expertise and effective organisation. It is only possible in democracies with the freedoms they offer to initiatives of this kind. Such democracies, however, constitute roughly two thirds of world governments, many among the richer and more powerful countries of the world. As well as reforming their own practices, they can also be pushed into promoting change in more autocratic and authoritarian regimes. Here we go on to consider how internal pressures within such countries themselves may cumulate and induce them to abandon climatically destructive measures and adopt positive ones.

2.3 Authoritarian policy-making within non-democracies

One third of world states are not democratic, in the sense of governments and their policies being ultimately chosen in free and fair elections. Many autocratic regimes do have regular popular votes, but these restrict popular choice either by allowing only approved candidates to be voted on (usually only members or supporters of the existing regime), eliminating popular opposition candidates by imprisoning or assassinating them, electoral fraud and control of communications and the media. Regimes claim the predictable results (sometimes 110 percent of the vote!), endorse their stay in power and the policies they are pursuing, but in practice elections have no chance of changing them. This rules out political party competition or government alternation as ways of inducing a more climate-friendly approach.

How then is policy decided? Here we must draw a distinction between highly centralised, technologically efficient, administratively effective regimes and others. These others are mostly under the rule of an army that simply represses opposition with force. Military regimes are found mostly throughout the Middle East with scattered examples in the rest of the world. Their main preoccupation is retaining power by any means and enriching their members. Ecological or climate concerns are hardly likely to figure unless pressure is put on them by other countries, most powerfully by their sponsors (which may include democracies in alliances such as the North Atlantic Treaty Organisation [NATO]). Military regimes tend to be highly dependent on these foreign protectors, so there is potential for exerting indirect pressure through the foreign policy of powerful democracies like the United States, if they in turn face internal pressures from their own electors to do so. To some extent too the heavy dependence of less developed countries on UN agencies and aid also opens up the way for internationally agreed standards to be enforced.

The major example of an authoritarian regime of the first type, economically advanced and administratively efficient, is of course China. Its enormous population and territorial extension, as well as its growing centrality to the world economy, make it a key player in any climate control initiative – and also the major contributor to carbon emissions. We have already pointed out that China has done more than any other country to limit emissions through checking population growth. Its one child policy from the 1960s to 2010s limited each family to only one child. It was enforced by brutal means (forced abortions for example) through police and pervasive Communist party surveillance. Highly intrusive, it was abandoned because of popular opposition and possibly growing prosperity, with women working full time at all levels and limiting family size spontaneously. In addition to the one-child policy, China has also developed zero carbon electrical generation to a point where it meets a major proportion of the national need (estimates vary from a quarter to a half). Once the Central Committee of the ruling Communist party has decided on a policy it has the means to pursue it with ruthless efficiency, driving it through, against whatever local opposition there may be.

The problem is that the Central Committee also has other interests and policy goals at heart. The major one is to pursue national economic development and growth at all costs, including climate costs. Thus, over half of energy production is still from highly polluting and carbon emitting brown coal; water is diverted and soaked up for industrial and agricultural use, rendering its sources increasingly salty; pesticides and chemicals are used indiscriminately on the land, eliminating insect pollinators entirely; whole forests are cut down in Siberia and South East Asia to meet Chinese demand; not to mention the incessant importation of Australian coal and mineral ore to feed factories.

Not only does the Central Committee itself have competing policy goals, with economic growth in first place, it also has many special interests embedded within the party structure itself, for example mining interests who stand to lose out comprehensively from carbon-limiting measures. The powerful position of local party officials means that schemes can be forced through locally even if out of line with top level objectives. Even though vigorous anti-corruption campaigns have been pursued in recent years, corruption remains endemic, permitting wide-ranging policy evasion and undermining measures that threaten imme-diate profits.

Being at heart a technocratic and bureaucratic regime, the main force for climate control measures to be taken are the scientists and experts of various kinds. Supported by the great polluted palls of fog hanging over Chinese cities, and prospects of dominating world markets in non-carbon energy production, experts have powerful support in arguing for emissions control. However, media clamp downs, censorship of any-thing resembling free political discussion and policy debate, the dom-ination of local party officials over any feedback that might reflect on their administration, all tend to weaken long-term scientific pressures for emissions regulation and policy change. Even specialists can be punished for disseminating inconvenient facts, as when doctors warned of danger at the outset of the Covid pandemic in 2019. Scientists and other experts clearly need support to strengthen their position, which may be supplied by the kind of pressures from outside that we consider in the next section.

2.4 Agreeing and enforcing policies internationally

The difficulty of providing public goods for the whole planet in the absence of an authoritative central government has already been outlined. The current international situation is like the situation within existing countries before their current state organisation formed and imposed its authority over its present territory. Purely voluntary arrangements to maintain the peace or initiate public works like roads or dams always ran into the free rider problem, where individuals who would enjoy the benefits anyway refused to cooperate or to pay. The only way was to force them to do so through an authoritative central state that could punish them for non-cooperation. Of course, states may operate at different levels with different kinds of government – national, regional, local – who may have competence over different territorial or policy areas. A good example of the latter is the Water State in the Netherlands, with precedence over all other bodies in maintaining the network of waterways, canals, dykes and the engineering that keeps half the country drained.

There are examples of similar functional bodies at world level, usu-ally to address recognised problems that cannot be resolved by states operating within their own territory. The best example is the World Health Organisation, which has some power to enforce emergency measures across the world. Its clout comes from fear of the immediate adverse consequences of its advice not being followed, for example the spread of the Covid-19 virus in 2019–2021. Were any territorial state to openly defy its recommendations it would probably be coerced by stronger neighbours or allies who fear the spread of the virus into their own territories, as well as by popular fears and unrest internally.

A useful way forward with regard to climate change and environ-ment would be to create functional bodies like the WHO, sponsored by the UN, in relatively uncontroversial areas that are clearly beneficial to everyone. The richer democracies might be willing to finance these due to internal campaigning pressures, while the benefits might even induce some authoritarian regimes to contribute. The most obvious example would be an Oceanic Pollution Organisation, whose imme-diate task would be to clear the canopy of plastic debris blocking the light and entering marine food chains in the middle of the Pacific Ocean, ultimately slowing and altering world ocean currents and wea-ther. Its activities could also extend to clearing plastics and chemicals from shorelines and enclosed seas. Success here would probably increase pressures for other functional UN bodies to be created, particularly as climate worsens and carbon capture through restored natural processes becomes ever more urgent.

There are, however, many specialised bodies already existing at world level that lack any immediate clout to enforce action on territorial states, as unfortunately does their parent body, the United Nations (UN) organisation itself. The UN indeed looks like a world government in waiting, with a legislature (the General Assembly) and an executive (the Security Council). Where both agree on action, policies can certainly be enforced, but any one of the five major countries on the Council can veto action, so it is easily aborted. Moreover, in many cases individual country governments can flout even internationally agreed policies without incurring any penalties.

There are of course examples in world history of individual states coming together to voluntarily link up and surrender some of their powers to a central government. This has happened quite often, mainly in the case of English-speaking countries (UK, US, Canada, Australia). Federations covering vast areas with previously separate states coming together as a proto-state and acquiring regulatory powers over most of a continent (the European Union), often have climate-related policies as a major focus. The UN looks rather like this but at a much earlier stage of formation. It is rarely able to coerce any of the 200 odd territorial states that constitute its membership and has to rely on voluntary negotiation and agreement among them – precisely the situation that national states emerged to resolve.

As a result, the focus for world agreement on tackling climate is on international conferences that try to negotiate agreements rather than impose them on the 200-odd state participants. Helped by growing evidence of the looming climate crisis, the Paris conference of 2015 was able to agree requirements for zero carbon emissions by 2050, requiring states to set and publish their own schedule for progress towards this over the intervening years. This was all enshrined in an international treaty, which thus entered as a requirement into domestic laws. This gave it some teeth in helping prevent domestic developments that failed to consider effects on emission targets.

The difficulties of relying on voluntary national compliance have, however, been sharply illustrated since Paris. Poland tried to modify EU requirements by postponing the zero-carbon date to 2070, while Australia and Brazil blocked progress at the Madrid climate conference of 2019 by claiming obsolete 'carbon credits' that would allow them to continue their current carbon increasing activities. Canada has carried on with a major environmental and climate catastrophe in the exploitation of the Athabasca oil sands in Alberta. Most damagingly, the US initiated procedures in 2018 to withdraw from the Agreement altogether on the basis of complete climate change denial, though these have been reversed with Democratic party control of the Presidency in 2021.

Intervening events may have worked to modify opposition to the Paris Agreement. The devastating forest fires of 2019–2021, increasingly erratic weather and flooding, rapid melting of the Polar ice caps and heating of oceans all demonstrate that climate change is happening faster than predicted. The relatively unrelated Corona pandemic of 2020 brought home to governments and public how easily and quickly global catastrophes can occur, destroying normal living in the process. The unprecedented economic interventions of governments prompted by the pandemic have undermined belief in autonomous markets and unregulated economic growth (a point pursued in the next chapter). All this may lend more weight, both in authoritarian and democratic regimes, to the growing scientific consensus in favour of climate action. However, this is always likely to be pushed aside in favour of immediate concerns like building up military forces or promoting national economic growth. Thus, we consider in our concluding section how pressure might be put on the laggards – including authoritarian regimes – by countries and other organisations convinced of the need for urgent measures. This can be done even within the fragmented international scene in the ways we now consider.

2.5 Getting climate action within the fragmented international system

Assuming we can win over public opinion and political parties in some of the more developed democracies (Chapters 3–6), how can these put pressure for action on the laggards? These include military based regimes whose main objective is to stay in power. If this involves capitulating to influential groups that ruthlessly exploit natural resources, and enriching themselves through paybacks and corruption at lower levels, that fits well with their main objectives. It has to be said, unfortunately, that many new democracies like Brazil and Russia also approximate to this situation, so the discussion of how to put international pressure on climate destroyers also applies to them, although with their somewhat greater internal freedoms there is more hope of working with domestic groups to limit climate damage. (The Polish ruling party has grudgingly, in late 2020, accepted 2050 as the date for stopping brown coal production because of internal opposition to any postponement.)

The other major case is China and regimes like it: technocratic, administratively efficient, but with an entrenched authoritarian government dominated by short-term goals like economic growth and international dominance. It also aims at complete political dominance internally, so any domestic discussion or perceived criticism of its policies are harshly and efficiently censored.

The saving feature of the regime from a climate point of view is that it is very technologically and scientifically oriented because of its developmental objectives. Hence scientific advice, particularly if unanimous, carries more weight than under less organised regimes (though always likely to be undermined by political considerations). Recent success in following expert advice on containing Covid has, on the one hand, strengthened experts' position, since the epidemic was worsened by suppressing early alarm calls from them, but it has also consolidated the regime's political position, given its relative success in containing the contagion. A promising recent development, however – a victory for medical specialists – has been the recently announced intention of abolishing the live food markets where the Covid-19 virus was possibly transmitted from animals to humans.

How can other countries and international bodies strengthen climate scientists' internal position in such a situation reinforcing the growing evidence for negative changes actually occurring? Creating a unified climate science community across the world with a common set of professional ethics is of course essential. These are closely linked to democratic values of transparency and free discussion. However, professionalisation is an autonomous process linked to world globalisation that can a) be left to itself and might even be compromised by being pushed politically and b) only works slowly, if steadily, to promote the action required by the climate emergency.

More immediate political action is made possible by the fact that any country aiming at growth and prosperity through exports is heavily dependent on its markets and the reactions of importers. The dominance of a free market philosophy at both global and national levels has made price practically the only consideration in international trade, in conformity with neo-liberal free market thinking. So cheap goods, from furniture to clothing, machines and electronics, have flooded world markets, undercutting domestic manufacturers more subject to safety rules and regulations, minimum wage requirements, inspectorates and quality controls, environmental restraints and prohibition of state subsidies.

International action against climate pollution could thus take the form of importer governments insisting that imported goods have to meet the same regulatory standards as those applied to domestic ones. Such standards could be adjusted to local conditions to take account of natural competitive advantages (lower cost of living justifying a lower minimum wage level), but such adjustments should not apply to environmental or health conditions. These requirements would also need to be backed up by on-the-spot inspections and checks and controls. Given endemic levels of corruption and administrative inefficiency in the exporting countries, as well as political interference and organised criminal gangs, inspectors would probably have to come from the importing countries. Controls could be reciprocal, of course, in that such inspection regimes could apply to all exports and imports anywhere.

As there is currently no world body to enforce such interventions, the penalty for refusal of foreign inspection would be embargoes by the importers on goods from non-regulated countries or higher tariffs on them, such as border carbon tariffs to prevent polluting production abroad penalising carbon regulated domestic industry. Such general measures are entirely practicable and, in fact, already there in the case of the EU and UK for non-environmental reasons. Other examples are primarily military-related as with US sanctions in a variety of fields against Russia, Iran, Cuba and Venezuela. There is no reason why other individual countries should not do the same. President Macron of France has already threatened an end to an EU free trade deal with Latin America if effective action is not taken against forest clearance in the Amazon region.

Foreign inspection would be entirely resisted by some regimes that would see themselves as being politically undermined by external intrusion. China is an obvious case in point. But if the relevant imports were then banned by importing countries, production in the most polluting sections of global industry would fall, being cut off from the export markets on which growth has depended. This would achieve the reductions required in emissions anyway and, by shifting the balance towards domestic industry in the importing country, lessen pollution from global transportation. A broader consequence would be strengthening the domestic scientific community's hand within authoritarian regimes, in arguing that the control measures they support from a scientific point of view must be applied credibly internally so that they can convince a sceptical foreign audience of their validity. Again, a practical precedent is provided by the imposition of US tariffs on Chinese goods in 2018–2019. Their success in obtaining some trade concessions shows that appropriate ones could be successful in strengthening environmental controls too.

A problem with less developed exporters like Indonesia, Brazil, and the sub-Sahara is that some of the clearance of rain forests is done by poor peasants on the fringes of state structures who operate in a haphazard way to clear plots for subsistence farming because they will starve otherwise. Neither the state administration nor a foreign inspectorate would be able to stop this effectively, precisely because it is so haphazard and disorganised. They might have more success with the large ranches and palm oil plantations now the main agents of clearance. The only clear and direct way to stop the peasant advance into forests is to provide an alternative – that is to provide them with a minimal income paid directly to them by the donors' cashiers, but only on condition that no clearance takes place in their local area or community. This could be verified by aerial inspection.

Direct payment to individuals themselves is necessary because of corruption, inefficiency and criminality in local administrations. Making direct individual payments and linking them to communal behaviour is less about getting poor peasants to organise armed resistance against the thugs employed by local landowners with a vested interest in land clearance – it is more to get them to organise politically, along with climate campaigners in cities, against the environmentally destructive regimes funded by landowners, mining and lumber corporations (while also stopping their own destructive activities of course).

Military and other authoritarian regimes, as well as corrupt democratic ones, are likely to organise opposition to these moves by presenting them as neo-colonial assaults on national sovereignty and stirring up nationalist fervour against them. It would be hard, however, for them to stand out against direct income payments to the poor, which would also stimulate the national economy and improve its tax base. The stick could also supplement the carrot, in the shape of the sanctions and tariffs discussed above. One aim would again be to increase popular support for climate experts and ecological movements internally who would gain more internal political clout and perhaps manage to take over the regime.

A further political problem, of course, is persuading the public and policy makers in richer countries to subsidise people in poorer ones. The more they are convinced of the imminence of a climate catastrophe (following on the recent experience of a pandemic and the massive government sponsored efforts to counter it) and the more credible individual foreign payments seem as an effective way of countering it, the more likely it is that such payments will gain support as a first step in combating climate change. It has recently been urged by the UN as a way of counteracting the social effects of the Covid pandemic. Massive sums are in fact already paid out by developed countries to avert impending problems, e.g. EU subsidies to Turkey to help keep potential immigrants within its borders and sums raised by charities for countries afflicted by wars and epidemics.

In the past foreign aid has been associated with immense industrial projects associated with environmental degradation rather than environmental improvement, with the money being siphoned off to elites in various ways. This has given rise to the apt definition of foreign aid as a way of transferring money from the poor of rich countries to the rich of poor countries. Controlling the flow of money to individuals should remove this stigma and even appeal to neo-liberals and free market economists who believe in individual freedom and enterprise. Distributing money more widely but in a better controlled way, is indeed likely to do more for enterprise and development (and avert climate change with its own costs) than any massive development aid put into the hands of the local administration.

2.6 General overview and conclusions

Everything would be so much easier if we only had effective world government! Public problems could then be resolved and effective solutions provided in the way they have been inside national states in the past. Even under the fragmented international system we have today, some action can be undertaken against climate change, though the absence of one controlling authority slows it down in the face of impending catastrophe. Mostly, action has to be taken within the boundaries of existing states and, for the most part, within democracies. Not only can these initiate effective measures within their own national boundaries, they can also back measures beyond their own boundaries by putting pressure on other countries to pursue them and supporting the domestic forces there that favour them. They can use their buying power, even in regard to technological giants like China, and much more within less developed countries open to interventions from outside. More imaginative ways of making foreign aid and sanctions work constructively are essential and not impossible, as current practices show.

In all this, democracies are key players in getting the necessary measures rolling both at home and internationally. A precondition for taking effective action to avert catastrophe is therefore to convince their electorates and governments of the need to do so. The chapters immediately following consider in detail how this can be done. First of all, climate campaigners and ecologists have to win the ideological battle with neo-liberalism, which puts national economic growth, private business and free markets above everything else (Chapter 3). Chapter 4 considers the rhetorical tactics needing to be employed to counter such attitudes within democratic political debate and the political alliances that follow on from this. Chapter 5 takes up a theme already stressed in the Preface – expanding support beyond relatively affluent climate activists to the masses of those semi-dispossessed (even in democracies) by globalisation and neo-liberal thinking. Just as with developing nation peasants, the democratic masses of developed nations need guarantees of basic security before they have the time and energy to support effective climate policies. This should help make democracies react more quickly than normal. A major stimulus will be if the middle classes, the main support base of current climate action, are more effectively organised to use their financial clout on governments. The last chapters revisit the international arena, to examine in detail how democracies mobilised by such domestic pressures can act more effectively there. Everything starts from winning the domestic ideological argument however, a point we go on to discuss now.

2.7 Further reading and reflection

Most of the electoral and democratic processes referred to here are analysed (generally, and without much reference to climate change as such) in *Politics: A Unified Introduction to How Democracy Works* (Ian Budge: Routledge 2019). Written as a textbook so as to be accessible to non-specialists and beginners, it brings in the latest research and findings on democratic parties, governments and institutions and processes in general. In chapter 15 it also deals with the international scene and prospects for World Government and World Democracy.

Green parties are steadily growing in strength across Europe, both at the level of the European Parliament and local elections. They gain fewer seats in national parliaments, however, and only rarely enter governments, let alone dominate them. 'Thinking globally and acting locally' is very important, particularly in mobilising public opinion and recruiting activists. Green parties and ecological movements have done a very good job here, but still lack real political clout. Where they might have gained it, as with the Five Star Movement in Italy, they have failed to make any really significant ecological impact. But perhaps we can hope for more of one now as they get buoyed up by the first grievous effects of climate disaster.

3   Dominating political discourse

3.1 Climate beliefs and their political effects

There are two kinds of discussion involved in the climate debate. The first is primarily scientific, centred on the question of what is really going on in terms of climate. The other is primarily political and centres on the question of what we should do about climate change while still maintaining something like our current socio-economic arrangements.

The two levels of discussion interact. If you believe that observed changes are within the normal range of variation for our geological period, then you will probably think we should carry on as normal. On the other hand, if you think the scientific theories predicting a heat explosion are increasingly borne out by evidence, you will want policies in place to deal with it.

There are also differences about the kind of climate change going on, between the idea that changes are gradually building up over an extended period of time and the growing perception that major changes are occurring now.

The long-term theory is the more attractive one for governments and politicians to adopt. It means that any fundamental measures to deal with climate change can be postponed for their successors to deal with, allowing them the freedom to concentrate on more pressing immediate matters – scandals, crises, wars, pandemics. The Covid crisis is a good case in point, leading to the postponement of the international climate conference (COP26) scheduled to be held in Glasgow at the end of 2020. Of course, the pandemic had to be confronted in early 2020, but (a) the virus was expected to be contained by summer and (b) precautionary methods had already been developed against it spreading at meetings (separated seating, virtual conferencing via TV and video links, etc.).

One reason for the postponement was the belief that when the Covid outbreak had been checked, hopefully by autumn, the overwhelming

DOI: 10.4324/9781003221630-3

priority would become a return to normal functioning of the economy. This would mean ramping up full manufacturing again, renewing international trade with global supply chains geared to last minute delivery as needed, massive tourism across the world, accelerated consumer buying and discarding of food, clothes and furniture, and enormous construction projects, all spurred by advertising across the media. These activities had mostly been suspended by governments during the pandemic in order to halt the spread of the virus. An unexpected side effect had been a fall in carbon emissions. Extinction Rebellion, Green parties, other ecological movements such as Greenpeace and even parliamentary commissions and expert advisers had already urged their immediate reduction as the main strategy against climate change – a position borne out by their beneficial environmental effects during the pandemic. It would have seemed appropriate therefore to proceed with an international assessment of what had been the most effective measures in reducing warming, balanced against their impact on social life and standards of living. Selective adoption of the most effective and least disruptive measures could then have been implemented on a considered basis with general world agreement, particularly in the rich countries that produce most emissions.

Instead, the main problem was seen as getting back to the pre-existing situation as soon as possible, abandoning most of the measures that had successfully reduced carbon pollution. This attitude is clearly summed up in the following quotation: 'the policies discouraging international transport for tourism or trade, shutting down a significant number of airline flights, altering our lifestyles to reduce economic activity, driving costs up and incomes down – all of it reversing economic growth into negative territory are ... regressive to the point of being anti-human ... making the vast majority of us poorer and costing lives as cash-strapped public services deteriorate' (Brian Monteith in *The Scotsman*, 23 March 2020, p 27). This admirably illustrates the ideological obstacles to action on climate and environment. Its major concern is with preserving high material living standards in rich countries, driven by privately-based economic growth at whatever environmental cost. It justifies private economic growth providing jobs (rather than massive environmental regeneration) as the only way to maintain a decent standard of living and keep public services such as health and welfare going (after a decade in which most countries have cut them in order to divert more resources to [private] economic growth!).

Scientific predictions about climate change hardly enter into this short-term, economically focussed perspective. Under the name of neoliberalism, however, it has dominated world political debate for the last

40 years. What the quotation underlines is that, besides beliefs about climate change, there is another intellectual force dominating discussions on what to do about it – free market economics. Most of its premises and its derived arguments are diametrically opposed to the policies that might avoid natural disasters. Panic reactions to the Covid-19 crisis have simply ignored conventional neo-liberal theory and (possibly) weakened it by borrowing and spending public money for social support on a colossal scale. Before we can reach any conclusions on whether neo-liberalism will return in full vigour, we have to examine in more detail what it is actually saying and how it impacts on climate and environmental considerations generally.

3.2 Neo-liberal economics

Economics has developed as a social science by generating general theories of the way free markets balance supply and demand for private goods through price adjustments. More goods and services will be produced as consumer demand grows, and then fall as it decreases. The less goods are produced the higher their price. As more are produced the price falls, as production costs per unit of output decrease. In a free market with many small producers, suppliers adjust production to price movements by making slightly more or less of the good available. Eventually supply and demand roughly match up at a price that gives producers enough of a profit to continue and keeps consumers buying at a rate that matches output. (Of course, there are many more nuances to the theory and there is not the space to develop these here, but they are described in the sources cited at the end of the chapter.)

Money to set up the production and distribution of goods is subject to much the same forces as the goods themselves, with its price, i.e. the rate of interest paid on the sums of money borrowed, going up or down with the demand for investment, which in turn derives from demand for the goods and services whose production will be financed by it. The stronger the demand for these, the stronger the demand for investment money (capital) and the higher the rate of interest. Wages and income are also determined in the same way. As demand for goods and services grows, more people are needed to supply them and wages and income increase – and conversely fall if demand and supply go down. In more extreme cases the workforce is cut to reduce the expenses of production and workers lose their jobs and incomes.

Aided by their ability to express these ideas about price, incomes, production, investment and employment in numerical, primarily money terms, economists were able to set them out in mathematical equations

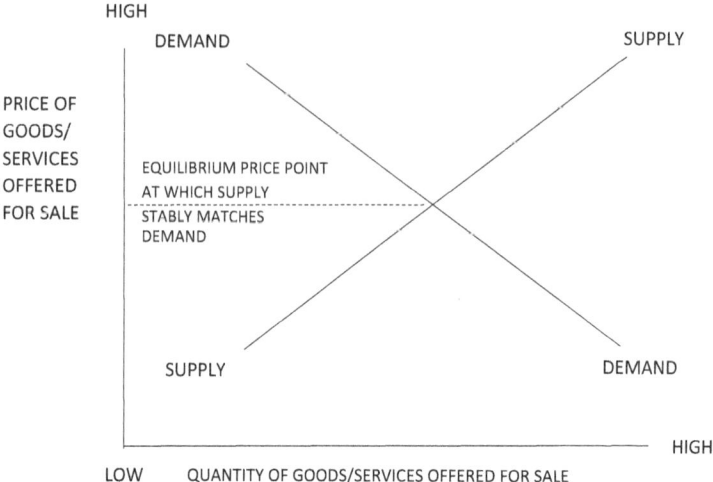

Figure 3.1 The relationship between the supply and demand of private goods and services in a free market.

akin to those of the natural sciences. Doing so made assertions about the market mechanisms involved more authoritative, almost like laws of nature independent of the particular social and political arrangements that produced them. The perceived need to strengthen free market mechanisms made these an intellectual and ideological basis for changing other social behaviours if they impeded the optimal results produced by their workings.

However, they also made the new economic science heavily dependent on the public statistics that showed how the market was performing – records of prices, goods and services produced, exports, imports, fluctuations in investment and interest rates. These were generally collected and published by state governments. As a result, economic science tended to focus on state economies, i.e. economic activities within the territories run by governments recognised by other governments as supervising collective and individual behaviour within their boundaries.

This was slightly paradoxical. Economies, instead of being defined on pure economic grounds as interrelated clusters of economic activities within a given area, or within a given production or service sector, were instead defined politically, by the boundaries of existing states. So, the major contexts or units of analysis for economics were defined as 'the British economy', 'the French', 'German' or 'US economy', primarily because all the available economic statistics were published by their governments.

National governments were also, of course, major economic actors within their own country, creating a huge demand for military and naval goods and for development of strategic railways, road, communications, etc., besides regulating general economic activities. Thus, thinking of states as the basic economic unit was not entirely unrealistic until globalisation created a world economy after the Second World War, linking up national goods and services to each other across the planet. This prompted attempts to standardise and aggregate national statistics so as to study regional and world economies.

In spite of such a broadening of concern, the main focus of economic analyses remains established states and territories defined by political boundaries. Within these the major interest is in the rate of economic growth – how far goods and services produced within state boundaries (the potential tax basis for government activities) have increased, or diminished, from year to year. For politicians and commentators alike the growth rate has become the major indicator of national virility and political success. All political debate tends to revolve around this, even in seemingly unrelated areas like health, education, military, environment and general well-being. This is because the assumption made in the newspaper quote above is generally, if not universally, shared – that growth is the only way of expanding such services or even keeping them going. Increased taxation or reallocation of internal resources are ruled out because of another widely shared assumption of free market economics – that taking money away from private businesses and economic enterprises to spend on public goods reduces the amount available for investment and hence reduces growth. That in turn is the only basis for maintaining public expenditure on all areas.

Breaking out from this narrow focus we might of course see other possibilities for financing public activity such as increasing taxes, particularly on those best able to bear them. Within the context of any one state this might be domestic business and the rich. Continuous private growth, however, is argued to be more of a win-win situation, where increased national wealth permits taxes to be lowered and public benefits to be financed while providing more jobs and income for more people. But to get into this happy situation one might have to increase productivity through labour-saving devices, which in turn require more private investment, provided by lowering taxes, with concomitant cuts to public services and demand, accompanied by higher unemployment. This is the situation that has prevailed across most countries of the world over the last decade. Vicious cycles seem as frequent as virtuous cycles under free market arrangements.

The fact that we now live in an integrated global economy rather than a collection of disparate country ones could change this reasoning. National growth is now tied to world growth, so that multinational corporations increasingly become the major players, along with the government, inside national economies. This opens up the possibility of increasing revenues to finance more public services by taxing multinationals more realistically, at least on revenues earned in the country – a proposal endorsed by the G7 summit meeting of rich countries in June 2021. At present companies operating in many countries take advantage of international loopholes to pay hardly any taxes to national governments, even if they benefit from the public goods these supply (education, health, transport, lighting, clean air, public order, security). Multinationals of this kind are mostly free riders of the type discussed in Chapter 2, who cannot be prevented from benefitting from the public goods provided but refuse to pay for them.

This is where state governments might come in, enforcing contributions from beneficiaries operating on their territory or imposing general taxes on them. At this point, however, multinationals and their neo-liberal apologists argue that extorting costs from such companies will reduce domestic investment. They will simply move to another country that taxes them less, so all the arguments against raising taxes on domestic companies apply with even more force to multinationals: taxing them for the public goods they consume will, in the end, lead to reducing these as the national tax base shrinks due to individual taxpayers' loss of income from multinational jobs.

This anti-tax argument can, however, be countered by pointing out that if a uniform tax is imposed on them in all countries, as proposed by the G7, there would be no advantage from moving. Besides, world companies are not completely mobile, being tied down to the countries they mainly operate in by domestic demand (and ability to pay) for the goods and services they provide. They also have infrastructural requirements such as good transport links, reasonable levels of security and (non) corruption, research facilities, administrative efficiency, healthcare, etc. So domestic taxation enters as a relatively minor consideration into their decisions about where to locate.

Up to June 2021 the need for 200-odd countries to agree on closing tax loopholes (as with measures against climate change) prevented any effective international action. Individual countries were concurrently blocked by their own neo-liberal parties and pressure groups from taxing multinationals realistically on their profits made domestically.

These groups are ideologically opposed to state taxation anyway and have deployed all of the arguments used above to thwart action on multinational taxation. The need for a stronger financial basis to confront the Covid pandemic has, however, pushed the seven richest countries in the world (the G7) into finally proposing to tax multinationals as described above, massively weakening neo-liberal arguments against it and providing a solid basis for climate campaigners to argue for it.

Neo-liberalism, as a political ideology, springs out of free market economics and uses many of its arguments, but it is not exactly the same thing. Free market economists can always argue, with some justice, that their science is focused on a general model of markets that, if its preconditions are met, would provide the optimal results for national economic growth. The qualifier is important, however, since its full preconditions are rarely met in actual practice. This is one reason advanced for the frequent inaccuracy of national growth and other economic forecasts. This failure is not therefore taken as discrediting the theoretical assumptions such forecasts are ultimately based on. The neo-liberal argument, now weakened, is rather that existing economic and other socio-political processes should be changed to bring them closer to free market assumptions, when everything would work better and be more predictable.

As noted, the major reform necessary within the free-market model would be substantial deregulation. Free markets would operate with minimum government interference (e.g. on environmental, health protection or wage controls) and low taxes, in turn leading to reduced public services. Where these do exist, they should, as far as possible, be contracted out to private companies rather than delivered by the public administration, as private competition leads to lower prices and greater efficiency in delivering the end results.

As in the wider free market, competition to provide public goods and services is driven by providers and producers seeking to maximise their profits by greater efficiency and cost cutting. In both cases these may carry external consequences – lowering wages, sacking workers, making work more intensive in order to increase productivity – and damaging the environment, e.g. by increasing emissions. Paying the costs of cleaning these up impedes free market operations and should be opposed. In the long run economic growth will provide the resources to pay for public goods, or more likely stimulate technological developments that may eliminate the need for them. But this can be left to the future – it is not an immediate problem.

Continued and unlimited growth of the national economy – as measured by the percentage increase in GDP this year compared to last – is, on these arguments, the real prize and to be achieved by any possible means. GDP is the total of goods and services produced within the country, less costs of production. This figure, published three monthly and annually, has immediate political repercussions. An increase in or above the 2–3 percent range means that governments can boast of substantial achievements in maintaining or improving growth, with the implication that it will benefit everyone. The fact that gains go mainly to the rich and that cuts in public services and contracting out to private companies artificially boosts the figure through various arcane accounting practices hardly counts.

National economic growth is the focus of political debate, but the fact that its measurement is based on a single figure causes problems in using it as the sole indicator of national well-being. For example, a mature wood may be cut down, boosting the income of the wood cutting company and possibly a timber company that sells on the timber, not to mention revenues from its later processing. Then for various reasons – perhaps to compensate for the loss of general amenities from the tree felling – new trees are planted on the site, stimulating even more economic activity for inclusion in the calculation of GDP. Indeed, the total of goods and services produced by the series of activities and services involved may be quite substantial – much better than if the mature wood had been left as it was in the first place (which would have been better from an amenity, environmental and climate perspective). From a general point of view, only monetary profits and prices count in the measure, so the more demolition and building gets done in cities, or the more high-speed railways are built to supplement existing networks, the better. Reducing travel through home working for example or by providing leave for baby care ultimately lowers GDP on present calculations.

3.3 Economics versus environment

Growth at all costs leads to environmental protections being weakened or abandoned, even against toxic gases still less climate changing ones. The contrast between economistic thinking centred on growth and environmentalist concerns with climate change emerges clearly from the summary of their assumptions and conclusions in Boxes 3.1 and 3.2. Box 3.1 simply summarises the neo-liberal arguments examined in the last section. Box 3.2 presents environmentalist thinking in the same way.

Of course, in considering government policy-making at all, you have to be clear what you are talking about. Both economists and environmentalists (Box 3.1, 2–5/ Box 3.2, 2–4) agree on policy-making

Box 3.1   Neo-liberal thinking about government policy-making

1. The major focus and context for government policy-making is the existing nation-state;
2. Policies are government decisions about what public goods should be provided for the nation-state, and how they should be delivered;
3. Public goods are products or services (including regulation or non-regulation of economic and other non-state activities) provided universally, so that nobody can be excluded from them;
4. These contrast with the private goods and services provided by free markets, which can only be acquired by individuals paying for them;
5. Public goods have to be paid for by compulsory general taxation, regardless of whether individuals want them or not;
6. Providing public goods therefore restricts individual freedom and choice, which are otherwise maximised by free markets (which allow individuals to choose what they want to pay for);
7. While some public goods are necessary for communal life (e.g. internal and external security), their limitation of individual freedom mean that they (and taxation to pay for them) should be kept to an absolute minimum;
8. Free markets guarantee the most efficient allocation of resources, in particular of capital investment in the production of goods and services, thus maximising national economic growth;
9. Private economic growth is the major contributor to individual and national wealth and hence to the government's ability to pay for public goods and services while raising individual incomes;
10. National economic growth is therefore the main public good governments should provide, overriding environmental and other considerations.

Box 3.2   Environmentalist thinking about government policy-making

1. Climate changes in any one country are affected by what is going on environmentally all round the planet, so that has to be the context within which all policy is made;
2. Policies are government decisions about what public goods should be provided and how they should be delivered both nationally and around the world (including what other national governments are doing);
3. Public goods are products or services (including regulation of economic and other non-state activities around the world) aimed at general benefits and provided universally, so that nobody can be excluded from them;
4. This contrasts with goods and services provided by private companies, which are only available to individuals who can pay for them;
5. Varying individual ability to pay for private goods creates massive inequalities across the world, which have to be evened out to prevent climate changing activities;
6. Public goods have to be paid for by more efficient government taxation both of richer individuals and large companies, including multinationals;
7. Providing public goods extends individual freedom by giving poorer individuals a more secure economic base, freedom and education to make their own choices (also about their environment) in a more informed way;
8. The major public good that governments should provide, overriding all other considerations, is reduction of the emissions that will produce abrupt rises in world temperature over the next few years with catastrophic social, political, economic and public health consequences.

as being broadly about providing public goods and services, i.e. those freely available to everyone in the population. These not only include tangible benefits such as security from domestic and foreign attack, but also services such as enforcing contracts and paying for judges and safety inspectors to enforce public safety regulations.

However, as other parts of the two Boxes make clear, the neoliberal and environmentalist evaluations of public goods provision and the benefits they provide then contrast very sharply. Neo-liberalism regards any restriction on the freedom of individual choice as in itself a bad thing (6, 7). It can only be justified (reluctantly) by measures taken to promote national economic growth (9–10), which include the security and law enforcement necessary to keep free markets going (8). Governments, acting within the territorial boundaries of the various world states (viewed as constituting the various national economies), should see their main job as promoting national development and growth rather than taking a global view of things (1). This then leads them back to their main task of promoting national economic growth rather than wider global interests.

Box 3.2 in contrast hammers home the obvious point that climate related concerns lead to a totally different set of conclusions, starting off from the idea that the main focus for governments, as well as the rest of us, has to be on the world developments that affect us and do not stop at the boundaries of any one nation-state (1). There is agreement with neo-liberals on what political discourse is directed at – government policies (2), and public goods (3) – and also on how these contrast with private goods (4). However, evaluations of public and private, and of their effects, differ sharply. As private goods have to be paid for by individuals, those who cannot pay lose out. This creates great and possibly growing inequalities of wealth within state populations everywhere (5), creating extreme poverty in most of the environmentally sensitive areas of the world. We have already noted this as one main driver of disastrous climate change. Providing free public goods from which they directly benefit (7) thus increases freedom in general by transferring surplus wealth from those who have less need for it to those who have more (6, 7). Ultimately, this also benefits taxpayers by saving them from the impending catastrophes that will hit everyone. Climate change threatens all our socio-economic arrangements (including free markets), along with the general well-being of citizens in all states and should therefore be the main concern of all their governments (8).

Like most summaries, Boxes 3.1 and 3.2 leave a lot out, but they are useful in terms of focusing the main arguments on both sides. It is important, of course, to realise that not everyone lines up at one extreme position or the other. In particular, a lot of people – and perhaps most governments – would broadly endorse the environmentalist position taken up in Box 3.2 (1, 8) about the need for a global perspective and for reducing emissions, while saying, in effect, 'yes, but not now'. It always seems more urgent to tackle immediate national problems, like the economic disruptions produced by the Covid pandemic, which threaten individuals' jobs and incomes, creating immediate problems with basic living standards and treats like holidays, flights and cruises. Yes, people say to themselves, we need to avert impending disaster, but let's get this sorted out first and then we can think about averting climate change in the run up to 2050.

It is precisely this attitude that plays into neo-liberal hands. They can present themselves as the pragmatists who will take the concrete action people and media are screaming for by restoring the economic growth that will provide all that they want, if not impeded by excessive government restrictions on carbon emissions from newly re-opened factories or on long ocean voyages and tourism. Governments, particularly democratic governments whose votes depend to some extent on responding to such popular feelings, are very inclined to adopt the neo-liberal argument and to let environmental concerns slip for a bit to get back to the good times there were before – with the best intentions, of course, to tackle climate change after they have got back there.

Doing so depends on accepting the assumption that privately based national economic growth is the only way to provide jobs and wealth and spend on public services. In principle, the environmentalist argument that these can be provided for the mass of the population by closing the numerous tax-avoiding strategies currently available both to the rich and to businesses, particularly multinationals, could be made equally if not more appealing to the mass of electors. The increased tax revenue would not be extracted from them and they would benefit from public services, including those averting climate disaster. These would provide jobs on a massive scale if tackled seriously. Such arguments have been increasingly made by some political parties over the last decade. Far from dominating the political debate, however, they have not been very successful at attracting voters or winning elections. In the next section we consider why, before going into more detail about it in Chapter 4.

3.4 The terms of political debate

The key to neo-liberal dominance of policy-making and debate is the fact that it focuses on the immediate problems weighing on the mass of the population – employment, income and jobs – and promises a concrete short-term solution to them in the shape of private economic growth. This will give everyone a boost, perhaps with extra goodies like cheap food, travel and (short) holiday breaks thrown in. These are all within the reach of hard-working families if growth can only be achieved. Even if sacrifices have to be made to get growth going, such as cutting public services for the less well off, the consequences for most individuals are not automatic and, in any case, lie in the future. In contrast the benefits from growth seem concrete and immediate – the next meal, the next holiday. Giving up on any of these to avoid climate change may have to be done later but not next week, next month or even next year.

Most of the media share this view, whether by conviction or default. Newspapers and radio stations are mostly owned by rich individuals or companies that subscribe to neo-liberal doctrines, support the political parties defined by them and frame political choices as ones between sensible, down to earth pragmatists, who will keep public expenditure under control and increase individual prosperity and freedom through free markets and growth, as opposed to ideologues who will sacrifice all this to avert vague threats in the future, which may not materialise anyway. Even the Covid pandemic and its accompanying economic lockdown is presented in this perspective as a temporary hiccup. Discussed at obsessive length, it also distracts attention from the floods, cyclones, fires and glacier melts that characterised the start of 2020. The tendency for short-term events to crowd out long-term concerns pushes even the media favouring palliative social and environmentalist measures into the short-term focus defined by neo-liberal assumptions. After all, if 80 or 90 percent of the national media are animatedly discussing a rise or fall in national GDP it is difficult for the others to give more than marginal or occasional consideration to long-term environmental or climate considerations, especially as they are usually bad news!

Concentration on private economic growth as the major political concern is reinforced by the fact that the main opponents of neo-liberalism in most democracies also see it as fundamental to achieving their own aims. Social Democrats and Labour parties favour social reforms and the reduction of inequalities, and hence greater provision of public goods and higher taxation to pay for them. They also support greater government intervention and regulation of economic life, e.g. on industrial safety and land use, in ways we shall examine more closely in the next chapter.

All this leaves them open to the criticism that they will hamper the free market, divert resources from individual and commercial development, and hence stunt the economic growth on which tax yields depend. This will generate a fiscal deficit that will undermine foreign investors' confidence, probably leading to them selling off holdings and thus to a collapse in the value of the national currency. This will hit everyone and undermine any social reforms introduced by Leftist governments.

The Leftist counter to such critiques is usually not to challenge the full set of assumptions on which they are based, but to promise to 'grow the economy' by direct intervention (e.g. by reinvigorating failing enterprises), putting money into research and technological development and improving industrial infrastructure. All this will increase GDP and pay for increased public services in the medium term.

Political disagreements are therefore focused on the means to achieve private economic growth rather than challenging the aim itself. Mild social reforms may also be proposed, such as a minimum wage requirement or higher unemployment and disability payments. To avoid accusations of profligacy, bias against business enterprises and higher taxation, the reformers again promise to 'grow the economy', thus surrendering to their opponents' premise that this is the sole way to achieve desirable social outcomes. This underlying consensus across the political spectrum on short-term, non or even anti-environmental objectives really seem to bar the way to any long-term climate action. Does this leave any room for political initiatives that might moderate the 'War against Nature' currently waged on all fronts?

3.5 Reframing the political debate? The impact of Covid-19

In contrast to the measures taken after the world financial crisis of 2008 (mainly 'quantitative easing' by pumping public money into banks and accompanying public service cuts), the need to quarantine much of the population to prevent mass transmission of the Covid-19 virus led to governments' closure of much of their industry and private businesses. Those able to work from homes did so. However, factories, shops, hotels, restaurants and transport needed workers on the premises, so they had to be shut down by government order to prevent transmission.

There were scattered protests by libertarians. Sweden did not introduce the general lockdown imposed elsewhere and the United States only did so patchily, but most governments enforced compulsory closures. These measures carried with them the threat of mass unemployment and loss of income. To compensate for this, governments made massive payments to employers to pass on to their workers, while also paying money in the form of social benefits to persons who missed out on these (self-employed workers for example).

All such measures drove a massive hole through normal financial practices – notably the neo-liberal concern with governments balancing the books by cutting public expenditures to match revenues, on the analogy of the household budget. Annual financial deficits rose to trillions, with no clear terms or times set for reducing them. Raising taxes was clearly not an option in the government induced recession. To avoid the financial panics that would have exacerbated the situation, the World

Bank and a consortium of national banks (and the European Union internally) guaranteed the loans to ensure that national governments could print more money and borrow freely to finance their unprecedented outlays. This was a situation that surpassed even wartime borrowing in the 1940s, when again the defence emergency required immediate unlimited expenditures that overrode all normal practices.

Despite vague hints about future tax rises, taxes on multinationals, and expenditure cuts to pay for it all, the scale of expenditures renders it unlikely the money can be paid off within any foreseeable future. It will simply swell national debts and the interest on them to massive amounts.

National debts were, however, designed precisely to push government debt repayment into the future. They are a financial device invented in the 18th century to pay for exceptional expenses, usually incurred in wars. Taking advantage of lenders who wished to find a safe haven for their money, governments issued bonds, i.e. written promises to repay money borrowed, usually in 20 or 30 years' time, while paying regular annual interest at an attractive rate (higher or lower depending on what alternative rates other borrowers were offering in the money market at that time).

Government bonds are a safer investment than company shares since there is less risk of governments collapsing or defaulting than other borrowers – they always have tax powers over their territory to draw on. Alternatively, they can sell off new bonds to raise money to pay off old ones. Generally, they can cover interest payments from current tax revenues. Lenders on their side can sell their bonds on to others in the money markets, just as they can do with shares in companies.

The main basis for these transactions is lenders' confidence that governments will be able to pay back the money borrowed when they said they would, and meanwhile to continue paying the agreed rate of interest on the bond. Confidence might be maintained by various means, one being balancing the books by cutting expenditures and public services. An alternative, however, might be having a credible plan to restore the economy, thus guaranteeing a stronger tax basis from the jobs and incomes this creates to pay the interest, and ensuring repayment of bonds to those wishing to cash them in. The paradox here is that few bondholders will want to cash them in if they seem safe and there are few other attractive alternatives. This is the case even though interest on the money is so low that inflation reduces real returns to below zero. So bondholders end up actually paying governments to take their money!

This surreal situation, in which investor confidence rather than any particular line of policy is what keeps financial crises at bay, is the aspect of public finance that modern monetary theory has seized on. Governments have unlimited credit if they can maintain potential lenders' confidence in getting their money back when they want it, and in the meantime can pay the agreed interest.

The important new point this introduces into economic thinking is that there is no preordained way to maintain financial confidence, particularly in the present unprecedented situation. Of course, if both lenders and borrowers are steeped in neo-liberal economics, and encouraged to do so by neo-liberal parties in legislatures and governments, they will see the way forward as cutting government expenditures as if they were somehow independent of surrounding social and natural conditions. However, Covid has to be seen – and *is* increasingly seen – not just as a health issue but also as an economic one. At the start of the 2020s it was *the* economic issue. Until transmission is overcome by the development of general immunity among the population, economies are not going to be able to function as normal.

Exactly the same arguments, of course, could apply to the growing climate disaster. Hurricanes, floods, heatwaves, devastating fires, are all economic dangers too. The more they knock out normal economic activity, the more they will undermine financial confidence, bringing in their wake financial crises as well as consequences for health and normal living.

Unlike Covid in 2020, there are increasingly accepted and simple scientific solutions to climate problems – cut human emissions and restore climate-calming natural processes. What is required are comprehensive government policies and actions to do so. Restoring the environment offers a solution to the post Covid economic crises by substituting for the 10–20 percent of economic activity that has disappeared as a result of the health measures. There is in effect no limit to the investment governments can put into reducing and absorbing climate changing emissions once their bondholders are convinced that such a programme is necessary – not only to get national economies going again but to anticipate the next natural and economic disaster in a way that secures their payments. The problem, of course, lies in convincing them. The multiplying natural signs cannot be ignored and are increasingly highlighted in the media and on the political agenda. Covid has opened the way to this. But now environmental campaigners must organise and develop new ways to hammer home the message in ways that put pressure on governments to act and get the general public on side. How to do so is discussed in the next chapters.

3.6 Further reading and reflection

An excellent summary of the 'Laws of Supply and Demand' is provided by Jim Chappelow in Investopedia (29 September 2019, https://www.investopedia.com/terms/l/law-of-supply-demand.asp). A succinct account of GDP appears in *The Economist* (www.economist.com), with a comment on problems with the measure, and others like GNP, associated with it (e.g. omission of the 'black' or unofficial economy and lack of concern with overall welfare as distinct from output). Such criticisms are amplified in a recent book by Dietrich Vollrath, *Fully Grown: Why a Stagnant Economy is a Sign of Success* (University of Chicago Press, 2019), particularly for attaching such importance to a single figure that ignores general well-being and other considerations such as ecological and climate costs. (But try telling that to politicians, even of the Left, except in New Zealand, which has already adopted a wider measure including environmental concerns.)

A succinct sketch of New Monetary Theory is provided by Iain MacWhirter in *The Herald on Sunday*, 30 August 2020, pp 1–2, (Facebook.com/heraldscotland.com). He also comments on the fallacies of basing all one's socioeconomic policies on a single dubious (and often politically influenced) statistic.

A comprehensive overview of Universal Basic Income, the many forms it could take and results from pilot schemes (e.g. in Finland) – and of actual practice where money has already been paid out generally on this basis (e.g. in the US state of Alaska) – is provided by John Lanchester in the *London Review of Books*, 18 July 2019, pp 5–8. This also gives useful references to general books on the subject. Readers will find a detailed discussion of an income guarantee at the end of Chapter 6 below.

An interesting point linking discussion of a basic income to the discussion of GDP and pursuit of economic growth is the finding that welfare and life satisfaction rise substantially as income increases from very low levels to a moderate one, but thereafter shows little correlation with further increases. See Jules Pretty, 'The Consumption of a Finite Planet' (Environmental Resource Economics DOI: 10.1007/s10640-013-9680-9 Springer Science and Business Media, Dordrecht 2013).

The necessity for restoring biodiversity to combat climate heating through complete environmental 'rewilding' is eloquently made and fully documented in David Attenborough's *A Life on Our Planet* (London, Witness Books 2020) pp 123–212.

4 Making climate the issue

4.1 The election setting

To work out in detail how the offer of an environmental economic initiative with supporting social measures would favour ecological movements and Green parties, we need to go further into voters' political thinking and the ways in which political parties shape it. A good starting point is the sheer difficulty (even for scientists, specialists, experts and politicians themselves) in deciding what are the right collective courses of action in the increasingly complex world we now live in. Ordinary citizens are asked to choose between policies such as going to war in the Middle East or relying on diplomacy. On the domestic side, they might be asked whether state support should be provided for healthcare through general taxation as opposed to taking out private insurance for it. Citizens are neither diplomats nor doctors, so how do they decide? This is even more true of climate related issues.

Public political debate is mostly about general and intangible policies, which may produce unanticipated consequences far in the future. Climate change is a major example of this, so pervasive and threatening that most people prefer not to think about it at all, or to switch off when they do. No wonder, when given a choice, they prefer a facile optimism about things working out in the end to drastic changes in their lifestyle that need to happen now. At least they know where they are in their current situation and for most it is not intolerable. Collective decision-making in most democracies is focussed on general elections, where the decision is more about the country choosing a government to run the country over the next 3–5 years than about policies as such. Of course, the policies that the parties forming the government will pursue are a consideration in voting for them, but other factors also enter in.

Popular votes are held from time to time in most countries to decide on particular policies (initiatives and referendums). These do provide

DOI: 10.4324/9781003221630-4

an opportunity to raise climate issues more directly with electors, but most policy still gets decided on the basis of the general parliamentary or presidential election outcome.

General elections are nominally concerned with electing individual candidates to parliaments to make decisions on behalf of their constituents. In practice, however, the candidates are almost always nominees of a political party – usually one operating throughout the country and directed by leadership teams ideologically committed to certain policies, who will participate in or run the next government if they get enough votes. The same applies to presidential elections. Electors may vote for one of the candidates on personal grounds or past record, but more important is which party nominates them. Having to make a choice of parties to form a government and act on your behalf makes it difficult to interpret general or presidential election voting in pure policy terms. However, choosing between party alternatives does simplify decisions for electors, as we shall see.

4.2 Party framing of public decisions

The difficulties of linking up what you want for yourself and your family with the public policies on offer are summarised in Figure 4.1, which illustrates the way voters decide on their vote choices in a general or presidential election. The central point this figure makes is that citizens need help in making the translation between the perfectly clear private aspirations they have for themselves and their family – centring on good service delivery and personal security (Box 1) – and the public alternatives on offer (Box 2). Help is provided by the political parties – directly, when they offer choices between general policy priorities in general elections (Box 3), and indirectly, when the governments they form make specific policy proposals for change between elections and implement them (Box 7). Both their election and inter-election proposals present voters with reasonably clear courses of action on public policy, which they can agree or disagree with.

There are thus two processes described in Figure 4.1 through which citizens can express public preferences: a process whereby electors react to the proposals continually being made by governments to alter existing policies in line with their overall programme (Box 7) and an election process through which parties collect votes and form governments, not only based on their policy programme but also the personal appeal of their candidates and their general record (Boxes 3, 4, 5). The different party programmes and government proposals and actions thus provide a concrete way for people to express agreement or disagreement with

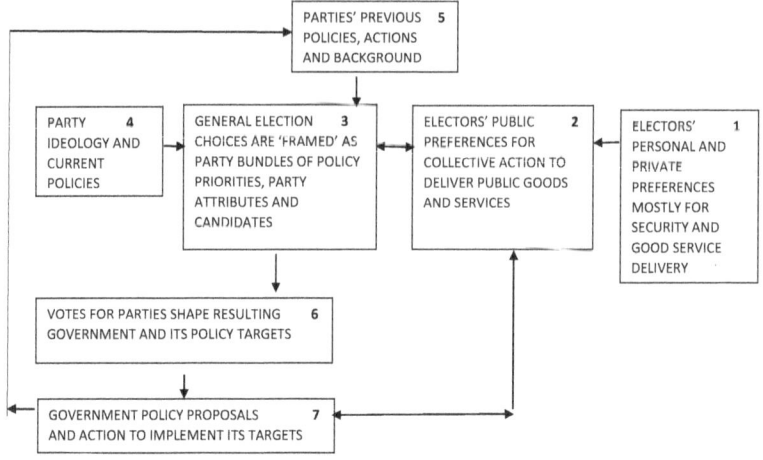

Figure 4.1　How electors' preferences are formed at election time and between elections.

policy priorities and their implementation (Boxes 3–7). Without parties this would be lacking.

Even so, voters often get confused between the policy alternatives on offer, so they give up on any detailed assessment of the issues at stake or of their consequences. Instead, they turn to 'decision aids' or shortcuts, means of assessing likely issue effects without going into detail about them. They may, for example, see all issues in terms of a general opposition between progressive forces and self-interested ones and hence vote for one party and its policy programme because it has a progressive record (or alternatively a history of defending their particular interests). These are the 'background factors' in Box 5 of Figure 4.1. An alternative, given the difficulty of figuring out long-term effects, might be to abandon any concern with policy issues and choose a party because it seems more united and competent in government or because the local candidate has more experience in the areas of interest to the individual voting.

4.3 How electors think about politics

Following from this we can see there are four ways electors can make choices about what party (and therefore what policies) they want to see in government:

a   They can use shortcuts to decide on a party that have little or nothing to do with its policy. These include its governing style, competence, experience and ability to get things done, or having likeable candidates. Deciding on this basis saves you having to make complex calculations about the future since you can just leave them to people you think would make the best decision under most circumstances.

b   Alternatively, you could take a policy-based shortcut by voting for a party in terms of its general policy orientation as shown in its past record. For example, if you were concerned with climate change you would vote for the party that had most emphasised this in the past and devoted a lot of its current manifesto or platform to the issues involved. This saves you thinking about the detail of any particular policy since it 'joins up' all of them as being for or against climate-related action. Thinking this way would lead you to vote for Green or ecological parties.

c   The most common way of 'joining up' issues, however, is in terms of Left versus Right, where the Left position includes promoting the welfare state through greater government intervention and fostering peaceful international co-operation, and the Right stresses freedom and the free market on the one hand and internal and external security on the other.

d   A fourth way for electors to choose between parties is on the basis of the particular issues that are important for them when they vote, e.g. if there is a problem with local flooding choose the party that seems most likely to focus on it, and similarly with a health problem like the Covid pandemic, and so on.

Of course, not all of the public will take the same approach to issues and policy all the time. They may choose on the basis of the most important issue at one time, but on a joined-up Left–Right basis at another. Their approach is likely to be influenced by whether an election campaign is going on or whether elections are not in prospect for another year or two. In an election period the political parties tend to locate all issues within an overarching framework (usually Left versus Right because it is easier to draw them together that way in their election programme). This then has an influence over the public's thinking, pushing them to see all the specific policy alternatives in the same overall terms.

Between elections, however, electors tend to be confronted by government proposals coming from separate ministries and debated separately in the legislature. Thus they start to see them as separate also, and react to them in their own terms and not as part of an overarching ideological policy framework. For example, floods can be seen as one-off or rare catastrophes to be tackled locally by better defences rather than as an inevitable consequence of world climate changes requiring a global response.

Electors may switch between these reactions at different times. This always has to be taken into account in political campaigning since some electors will always think in separate, issue-by-issue terms and some-times a majority will look at things in this way. Political campaigning thus becomes a question of making your issue(s) more important for electors than those associated with rival parties. Usually, it is the short-term issues with immediate effects that predominate over the others.

These reactions suggest that a firm plan for environmentally driven economic recovery, supplemented by a guarantee of a decent income for all, should always be the major plank in Green and ecological parties' platforms. These are single issues with an immediate appeal and imme-diate effects. Such initiatives do not need to be fitted into a broader argument about climate change to make them appealing. They are attractive as single policies on their own terms – improving life immedi-ately for some and providing future security and present peace of mind for the majority. These also have a practical recent precedent in the special measures and income support provided by most governments during the pandemic shutdown.

The strength of the income guarantee as a political issue is that it can also be fitted into a wider policy framework for voters who think in such joined up climate terms. It can be seen as necessary compensation ensuring a 'just transition' for those adversely affected by the other cli-mate change measures like rewilding, creating marine reserves, cutting pollution and so on. While fitting well into this overall perspective, it can also exert a powerful appeal on its own, however.

Climate activists – often coming from a Leftist ideological perspec-tive themselves – tend to see all of politics in joined-up terms. Hence their political campaigning aims at converting electors to a full-scale realisation of climate perils and their causes. But they have to realise that this may put off many who think in single issue terms. The time and effort put into expounding the whole climate change scenario to every elector may thus not be an efficient use of resources and time. That in turn may explain why climate campaigners and their parties have made only limited vote gains so far, making them junior parties in governments and mere adjuncts in Left dominated election alliances. Since parties frame and shape electoral choices, the Green failure to assert themselves in more immediate terms has major consequences for the votes they can attract and hence the policies they can promote. We therefore go on to examine parties, their interrelationships and current standing in the next section.

4.4 Political parties as main agents of policy change

Modern states have all developed the institutions – army, bureaucracy, law courts, police – without which they could not defend and administer their territory. The most basic distinction between states, however, is whether public policy is decided exclusively by the military-bureaucratic elite, alone and apart from the rest of the population, or whether other institutions have been added to express and impose popular preferences on them. The most important of these are regular competitive elections, freely formed parties, an independent media, parliaments determined by election results and governments responsive to such parliaments.

Central to such institutions are the political parties, since they operate at all these levels and link them up – framing choices for voters, focusing candidates on a common programme in elections, organising them to support it afterwards and forming governments to carry it through.

Mass parties were originally organised by social groups (workers, farmers, religious groups, cultural minorities) that felt themselves politically excluded under the existing set-up. More established groupings, the landowners, business entrepreneurs and established churches, had to follow suit in order to defend their interests. By asserting their claim to direct state policy in line with the views of their supporters, and then competing in free elections for control of policy-making, mass parties rendered the population pro-active in making the political decisions that affected them, rather than having to simply submit to policies made by their rulers.

The mass party transformed the political situation in three ways. They created state-wide organisations that grouped supporters in local branches. Such branches collected money to finance party activities, giving the party a financial base independent of state subsidies. They also rallied and organised supporters in elections contested between parties in order to get the maximum number of party candidates elected to the legislature and thus empowered to push party policy there. To do this they developed ideologies – extended socioeconomic and political analyses of the national and world situation – that justified the policies they were pursuing in the interests of their supporters. A good example is the neo-liberal ideology (see Box 3.1 of Chapter 3) justifying private business expansion above all other social considerations. The main opposition to this has traditionally come from Socialist and

LOCATING POLITICAL PARTIES FROM LEFT TO RIGHT ALONG A LINE

Greens	Social democrats Labour	Christian democrats Other religious parties	Liberals	Conservative parties
Communists Left Socialists	Some Farmers' parties	Some Minority Ethnic parties	Some Minority Ethnic Parties	Some Farmers' parties	(State) nationalist parties

Left

Right

Stress on welfare and government intervention to secure it, and peace.

Stress on freedom, traditional hierarchies and national security.

Figure 4.2 Locating political parties from left to right along a line.

Labour parties pressing for worker protection and welfare with greater state intervention to get it.

We illustrate how the parties place themselves (and are seen by electors) in such Left–Right terms in Figure 4.2. Here we see that Greens are grouped (and group themselves) very much with the Left, owing to their tendency to take up the same social issues as Socialist and Labour parties. This can be justified on the grounds that these parties are natural progressive allies who see nothing wrong with government intervention for the public good. Green support can push them into a more radical pro-environmental and pro-climate stance. Taking up other progressive causes, however, blurs the Green focus on climate, which is an issue that could appeal across the whole political spectrum. This is especially the case for some party families like agrarians, minority nationalists and above all Christian democrats. Christians put their major focus on the maintenance of the Christian family, do not have very clear Left–Right positions and might be wooed by Greens if their parties were less firmly positioned on the Left.

Minority ethnic parties arose alongside class-based parties in the 19th century. With their concerns focused on decentralisation of powers or even political independence for their own region, they are even harder than religious parties to classify in Left–Right terms, as are agrarians representing the interests of farmers both small and large. So they are open to other lines of persuasion in terms of climate change.

Parties' characteristic issue concerns form the starting point for examining their electoral appeal. As pointed out above, many electors see the world in single issue terms and vote for the party most clearly associated with an issue that concerns them. That issue may change from time to time between different elections, but if it is clearly associated with a particular party, that is the one they are going to vote for in that election. The strategy on the party side must be to make their issue the one of most concern to most voters. As already pointed out this may be difficult to do with climate change, which demands a long-term perspective and acceptance of new regulatory burdens rather than immediate rewards. That is the reason for stressing an income guarantee as one of the core issues that Green parties should emphasise, along with environmental restoration to provide jobs. The latter can be presented as the solution to shattered economies, providing growth and jobs after the pandemic, while the income guarantee provides economic security for everyone.

Of course, there are other voters who do think in joined-up terms about politics, particularly at general elections when the parties themselves tend to present the issues in this way to simplify the many issue choices involved. The problem for the ecological parties is that the joined-up choice generally offered is not one of climate disaster versus short-term economic growth. What dominates is rather the traditional opposition of Left versus Right.

Most political comment at election time is thus about the extent to which the parties are 'moderates' or 'extremists' – will move to 'left' or 'right' or 'capture the centre ground' (already occupied by such parties as the Christian democrats with a mix of 'left' and 'right' priorities and some of their own religious traditions). It is as if political parties, in stating their priorities, took up some position on a line between a pure Left and a pure Right position and moved from side to side along this at different elections. Following on this lead, many electors and voters also describe themselves (or others!) as Left or Right and, on this basis, they choose to support one party rather than another. The pervasiveness and power of such thinking means that Green and ecological parties, rather than being viewed in their own issue terms, are usually evaluated in terms of traditional Left or Right concerns with their own being totally passed over. What is most obvious from Figure 4.2 is the way in which Greens are generally seen as an extreme party of the Left – more so even than Social Democrats and Labour. That is partly because stopping climate change involves more governmental intervention and regulation, which Left-wing social reforms would also require. But it also stems from the Greens' own espousal of so many other progressive causes such as women's liberation, pacifism, the European Union, nuclear disarmament, and so on. Many of these can be linked to stopping climate change, but they are not central to it and they are all generally Leftist in their orientation. This makes Green parties captives of the other Left parties, rather than pushing them into making climate change the priority.

In contrast, other party groupings outside the main Left–Right division, such as agrarian parties, have avoided being so easily stereotyped in its terms, and ethnic minority parties are placed, if at all, at different points on the Left–Right line shown in Figure 4.2. This leaves them freer to pursue their own – and their supporters' – particular concerns with devolution or independence and to form alliances with parties anywhere along the Left–Right spectrum. The lesson for Green parties is therefore to promote an alternative form of 'joined-up' political thinking for those who view politics in this way, while reinforcing their appeal to those who decide on the basis of single short-term issues by offering immediate pay-offs from environmental job creation and a minimum income guarantee. We go into each of these below.

4.5 Getting away from 'politics as normal'

Democracies are the key agents in the fight to combat climate change because they tolerate the popular actions that can push their government into doing so. However, ecological parties and movements have mostly failed to make much of an impact, remaining minor players in elections, parliaments, government and policy making – even policies on climate change. If campaigners cannot carry their politicians and publics along with them under the relatively free and open environment that democracies offer then there is little hope of doing so or exerting pressure under more repressive regimes at governmental and electoral levels.

Part of their failure can be attributed to a lack of charismatic leaders with fire in their bellies who can develop a general populist appeal. Coming predominantly from the professional and scientific classes, Green politicians are, by nature, usually reasonable and prepared to accept compromises, even in the face of impending catastrophe. They have therefore been willing to enter into the games of politics as usual and to allow their main message to be diluted with the other issues prioritised by their partners and opponents, such as financial crises, austerity, economic growth, Brexit, resurgent cold wars, pandemics, Black Power and feminist movements.

Certainly, some of these crises set precedents that Green movements might draw upon for the urgent actions needing to be taken on climate. But they are issues that other parties have already pre-empted, and that, as we have seen, are as often used as arguments for downgrading climate change in the face of other problems as for prioritising it.

Green parties cannot afford to play second fiddle any longer. To make an electoral breakthrough they really need to emulate the populist parties that erupted on the democratic scene in the late 2010s. There were two elements to the populist appeal. The first was an unconventional leader challenging established procedures, parties and politicians and promising to replace them (Le Pen, Farage, Salvini, Grillo, Trump). As we have noted, voters who find politics confusing often ignore issues and decide their votes on other grounds such as personality and impressions of competence. Populist leaders aim at providing these in bags, while attacking the existing 'political class' for ignoring the people.

The other element in populist successes has been their concentration on a single, easily understood issue that linked up immediately with widespread popular grievances. This is usually mass immigration from the underdeveloped world, perceived as threatening jobs, housing, welfare and, more generally, national identity. The established 'political class' was then portrayed as selling out the native population by being soft on the incomers.

Populism has been successfully opposed by those established leaders who have emulated its style, first differentiating themselves by having broken into the establishment rather than coming up through the usual channels (Macron, Johnson, Obama). Then by having one big issue to hammer – usually promises to 'get things moving again' – centring on economic growth, but supplemented by immediate benefits from healthcare (Obama) or Brexit (Johnson). The combination of personality and a single driving issue seems effective in rallying voters in the cases both of populist and established parties that steal their thunder.

Why should climate ignorers have the best appeals? Actually, climate campaigners have shown what more could be done politically with Extinction Rebellion. This has pushed alternative modes of political action further and challenged conventional political concerns. It has also produced a charismatic and unconventional leader in the shape of the teenage Swedish schoolgirl Greta Thunberg. Unfortunately, and in a way typical of professionals and idealistic Green campaigners, Extinction Rebellion has contented itself with purely symbolic and/ or rhetorical victories – occupying city streets for a week, or publicly berating world leaders. It has not gone for the jugular of electoral politics by having national leaders in the mould of Greta denouncing every failing of the establishment, perhaps using Covid failures as a springboard, and hammering home the imminence of disaster and need for personal security through environmental and job creation and a minimum income guarantee. These single issues would powerfully supplement personal attacks on establishment failures to provide for ordinary people while taking away livelihoods through a quarantine that should not have been necessary. There is unlimited ammunition here for those who care to mine it rather than being always Mr(s) Nice Guy. Why should the devil have the best invective? Electors frightened by

Covid-19, and other disasters to follow, are prime material to mobilise in the face of climate change and economic disruption.

But they require immediate payoffs – or the promise of them – to be properly mobilised. This can be found in a massive environmental programme to 'get the economy moving again' and avert climate disaster and in the single-issue income guarantee that provides security against all future disasters. The advantage is that, as well as being immediately appealing in their own terms, these issues also tie in with a long-term coherent programme for tackling world climate change. With its basis in scientific theorising and increasingly confirmed predictions, the advocacy of such immediate social and economic reforms also fits into a joined-up climate-based perspective that has the capacity to replace Left–Right as the dominant framework for analysing political problems.

Green parties should campaign on both fronts, to draw in voters who think in joined-up as well as single issue terms about what is at stake. This might also extend Green appeal further away from the Left to voters for the centre parties, Christians and agrarians, whose outlook renders them more open to planetary concerns than neo-liberals and Conservatives. These concerns should be sharpened by the continuing escalation of the climate crisis in the 2020s with continuing floods, storms, rising sea levels, murderous heat waves and polluting fires.

The central question, however, is whether Green parties and movements are capable of turning these events into political capital by dominating political discourse, winning elections and leading governments. Reforming themselves and changing their appeal should help. We look at the detail of doing so in the next two chapters.

4.6　Further reading and reflection

For an extended analysis of electors' political thinking and the way it is framed by political parties, see Budge, *Politics: A Unified Introduction to How Democracy Works* (Routledge 2019) chapters 3–5 and 11. The important point is the continuing interaction between parties, who present and frame the issues to be decided by election results, and the voters who make the final choices, but often on non-policy grounds or on very short-term considerations to which parties must then respond. How to get political action in democracies to avert climate change under these conditions is what we consider in the next two chapters.

5 Strengthening mass appeal
Guaranteeing a decent income for all

5.1 Creating a winning electoral appeal

Normal democratic processes are generally too slow to produce the urgent action needed to avert climate change. However, they have been drastically accelerated by the measures taken to confront the Covid-19 virus and the socio-economic emergency it triggered. Such policies would have been ruled out as impossible before the pandemic struck. They are still ruled out as permanent elements of the 'new normal', but the emergency has demonstrated that vast sums can be raised and spent by governments on keeping businesses going, supporting incomes and providing services without triggering a financial crisis. Indeed, these emergency actions seem to demonstrate that New Monetary Theory is correct in characterising state credit as practically limitless, provided it maintains lenders' confidence and trust.

These are not going to be maintained, however, in the face of business collapse and mass unemployment following on from the quarantines and restrictions imposed by governments to stop the spread of Covid-19. Combating unemployment and recession by conventional neo-liberal methods of 'quantitative easing' (boosting bank and business capital by buying up worthless assets), while savagely cutting public services to 'balance the books' (all representing transfers of wealth from poor to rich) are just going to extend the crisis. Resulting inequalities will destabilise the situation further by fuelling political discontent. Propping up failing traditional industries like steelworks or shipbuilding is not going to work either, nor are subsidies to the sugar puff fringes of mass tourism, travel and catering, all of which are going to shrink under the impact of Covid and successor pandemics. Building or subsidising infrastructure to support these activities (e.g. new roads, airports, harbours and railways) is pointless when internal travel has shrunk as a result of staff working from home. Subsidising construction of poor-quality

DOI: 10.4324/9781003221630-5

housing by relaxing controls is also unlikely to impress bondholders when the white-collar workers able to buy them are increasingly able to relocate themselves in better houses in remoter regions.

'Austerity' – aka public service cuts creating private wealth amidst public squalor – will no longer be enough to prop up financial confidence and promote national economic growth even to the delayed and limited extent it did after the world financial crisis of 2008. Instead, Covid, and expectations of future pandemics of the same kind, have underlined the need to enhance public services that can cope with them. The looming climate disaster ever more evident around us will increase such pressures.

In the face of all this, the only viable solution to economic collapse and financial loss of confidence will be for governments to tackle these problems head on, by financing public services and climate-ameliorating initiatives. These will simultaneously create new jobs and boost economic growth in a socially useful way. To compensate for the jobs lost in current climate damaging activities (e.g. in commercial fishing to extinction), the new jobs from environmental initiatives need to be supplemented by a minimum income guarantee. This will not only cushion job losses but also provide automatic compensation – free from cumbersome and protracted bureaucratic procedures – for the flooding and destruction following on from climatic disasters. Massive environmental job creation and the minimum income guarantee both have great potential appeal for voters and electors. Any party or political movement picking them up and running with them – particularly if hammered home in a populist style – singling out the 'guilty men' bringing about disaster in the first place and then profiting from it – is likely to sweep up votes. Even if not ejected from office, the existing parties will be forced in sheer self-defence to take over these policies themselves.

The most likely organisations on which to base a mass environmental populist party are Extinction Rebellion and Greenpeace. (Green parties are mostly too embedded in existing procedures and institutions to take up populist crusades and are already doing useful work in raising climate consciousness within local governments and organisations.) Of the two, Greenpeace has been established longer. Its forte has been to undertake bold actions against environmentally threatening practices such as Russian oil drilling in the Arctic Ocean, Japanese whaling in the Antarctic, experimentation with genetically modified crops in Britain and illegal fishing in conservation areas in the North Sea. The actions have been small scale and symbolic, designed to draw attention to threatening practices rather than mounting large-scale campaigns to stop them in their tracks.

Extinction Rebellion on the other hand has organised a systematic mass campaign of school strikes across Europe and, to a lesser extent, in North America, various mass demonstrations and an illegal but non-violent occupations of central London and other cities for a week during the summer of 2019. It has also promoted headline-grabbing confrontations between the figures of Greta Thunberg, an articulate and assured Swedish teenager, and various world leaders accused of hostility or apathy about climate change.

Both these organisations are committed to non-violent (if sometimes mildly illegal) and sporadic action, designed to push climate change up the political agenda, trusting that established political leaders will then take effective action against it. (Not a hope that has been fulfilled, given the latter's concentration on short-term issues.)

One feature of Extinction Rebellion is their distrust of any hierarchical organisation, preferring spontaneous cooperation between self-organising small groups. That, at any rate, was a major feature of the London street occupation of 2019. However, there was clearly an effective organisation promoting Thunberg's travels around the world and appearances at the United Nations and world summit meetings, followed by crowds of enthusiastic supporters.

Equality, consensus and a distrust of hierarchy are features of these two movements shared with 350.org, a widespread American environmental network also geared to promoting environmental causes. Mostly progressives of the Left, they tend to believe that the case for climate action is so reasonable and overwhelming that you only have to bring it to general attention through striking actions and demonstrations, supplemented by computer networking, to make it the focus of political action, brushing aside the petty concerns of the moment.

Unfortunately, these are precisely what do preoccupy established parties and politicians, starting with getting re-elected and retaining power and, secondly, tailoring public policy to their ideological concerns, which rarely prioritise climate change. Indeed, as with neoliberalism, they may even promote it. Even the looming prospect of new pandemics is neglected in favour of finding immediate solutions, hopefully a vaccine, to the existing Covid-19. National discussion and action are focused on the immediate problems connected with it, particularly the economic and social effects of quarantines and lockdowns and the adequacy of government responses, rather than long-term prevention of pandemics in general.

Green parties and their leaders have, in contrast to extra-parliamentary campaigning movements, realised that continuing, organised political effort is needed to get real action. However, participating in conventional ways in conventional politics has snarled them up in their concerns, which are short-term and immediate rather than the radical re-orientation of policies needed to avert impending disaster.

The shocking impact of Covid-19 and the panic measures taken to minimise its consequences certainly open the way to drastic action, but are climate campaigners and their representatives capable of seizing the opportunity? One has to say, not in their current shape. The one chance is for the campaigning movements to realise that sporadic and largely symbolic action is not enough. They need to infuse an organisational apparatus more socially extended than that of Green parties with their own evangelistic and populist zeal, attacking all existing parties whether established ones or their current populist challengers, fighting elections on issues that provide an immediate payoff for distressed and disorientated voters and which also promote climate calming measures. These can all be combined in the environmental New Deal with its potential for job creation and the minimum income guarantee, both discussed below. These should start hurting governments where it matters to them, in terms of votes. A parallel campaign to hit them even harder in terms of their tax base is considered in Chapter 6. Organising politically on a continuing basis and introducing innovative forms of political action is, however, the only way to avert disaster. If not now, then never. And soon the heat bomb will explode.

5.2 Winning elections

Up to now Green parties and ecological movements have mostly failed to win elections or, where winning outright is uncommon, even to make themselves the largest (plurality) party. The one exception is the Italian Cinque Stelle (Five Star) movement, which did receive around a third of the national vote in the general election of 2018. Its success stemmed partly from widespread disillusion with the existing set up and from a policy of support for the unemployed that made the payments system more generous. It was also helped by the political reforms of the 1990s that moved the electoral system closer to proportional representation (PR), a way of aggregating votes that aims at making party seats in the legislature directly reflect vote shares. This allows new parties to make their mark quickly instead of having to get over a high threshold before obtaining representation, as under the one-member constituency-based system of most Anglo-Saxon countries.

The Cinque Stelle was strongly ecological in its orientation but also espoused many other social and political causes, so it might be regarded more as a generally radical party than a purely ecological one. In any case, its position was weakened by the success of an equally populist anti-immigration party in the same election, which was the only one it proved politically possible to form a coalition government with. The period of weak and divided government that followed discredited the movement, whose local administrators also failed to deal effectively with pressing ecological problems. Eventually it was forced into a coalition with established parties and lost its political momentum.

Other Green parties have succeeded in making themselves pivotal in coalitions of the Left, notably in Germany round about 2000, having first campaigned on a mix of ecological and social reform issues. These may even make it the largest party in the current (at time of writing) federal elections, though taking over the government is likely to be blocked by a coalition of the established parties. Under Green influence, Germany has certainly taken a lead in palliative ecological measures, but hardly the breakthrough necessary to combat immediate climate change. One explanation for Green underperformance when its representatives got into governments in Italy and Germany, as well as in other countries, is overextension to a wide range of social issues, mostly shared with parties on the Left. This has, in many cases, led to Red-Green electoral alliances espousing a wide range of social and political reforms. Some of these certainly link up with climate change, e.g. general empowerment of women. However, the particular issues involved are often those affecting educated Western women rather than the really downtrodden in the developing world, where even limited emancipation would radically change things (birth rates for example).

Overextension to a variety of issues besides those linked directly to climate's existential threats is certainly one reason for relative electoral failure. Another is the Greens' strong link, or near identification, with the Left. This is natural given that most Green activists originate from Left-wing parties that are more receptive to proposals for radical reform in all sorts of areas and support wide-ranging state intervention to do this. This makes it easier to form coalition governments or electoral alliances with them. The problem, however, is that such governments want to deal with the more pressing social questions first (people are dying after all). Moreover, Left parties often link new social problems to the unemployment and poverty produced by the decline of old industries and industrial regions, so their priority is often to combat deindustrialisation and strengthen economic growth in these areas, which leads to a position not unlike neo-liberals in the UK. Growth must be fostered to provide jobs and income, after which we can worry about emissions and heat reducing measures. Red-Green alliances thus make Greens the prisoners of a Leftist social agenda rather than prioritising a climate calming one. This has two consequences:

a   It disperses their energies and appeals over a wide range of issues, many of which have little to do directly with climate change. They then get sucked into politics as usual and a myriad of day-to-day concerns, often procedural ones, rather than the existential threat of the developing heat bomb.
b   This is doubly unfortunate because the threat is one all parties and their supporters from all over the political spectrum face, if only they can be convinced of it. The Covid emergency offers a chance of doing so, but Greens lining up so decisively with the Left puts other parties off.

Neo-liberal and Rightist nationalists may deny the immediate climate threat but centre parties, particularly Christian Democrats, are easier to win over after Covid. If they see climate change as essentially a Leftist cause their conversion to immediate measures against it becomes harder. A mass environmentalist party should aim at potential alliances and coalitions across the entire political spectrum by focusing ruthlessly on climate change and immediate measures against it. The most important of these are coherent environmental initiatives undertaken nationally to reduce emissions and build up carbon-absorbing natural processes and a universal minimum income guarantee. Both have the additional merit of providing jobs and income to avert immediate economic collapse and convince bondholders and investors that there is a relevant long-term plan in place to tackle impending problems. For environmentalist parties that adopt them as their causes, these are single issue appeals that tackle the immediate concrete problems afflicting the mass of voters. What's not to like about credible programmes to provide jobs and adequate sustenance in the face of disaster? Kicking out the rascals who provoked it in the first place adds an additional flavour.

To make such issues work for them, however, climate campaigners have to ally, if not combine, and organise themselves efficiently. A first step would be to call a summit conference to set up the necessary national political network. Forcing through amalgamations with the other climate-oriented movements could be rancorous and self-defeating, but at least electoral alliances could be agreed and some kind of division of political labour decided, with a coordinating council to avoid unnecessary and vote-wasting competition. This would vary by country, of course, but some kind of evangelistic populist mass movement is needed to add the necessary impetus to the climate cause. And it has to be organised from national down to local level to follow up its appeal, with continuing mobilisation of all age groups through social and educational as well as political activities.

5.3 Election tactics

Concentrating on out-groups who are clearly distinguishable and identifying them as the cause rather than the effect of global changes has proved a winning election tactic for nationalistic parties since the mass migrations of Blacks, Irish, Poles, Muslims, Jews, Italians and Roma in earlier centuries. It offers a simple explanation for all the social ills affecting the native population and provides a simple solution: stop immigration. It is also a diagnosis shared by elements of both Left and Right: industrial workers see working conditions and wages undermined by cheap immigrant labour while small shopkeepers and the lower middle class in general face new competition and intrusion from the newcomers. As an issue, therefore, it can disrupt traditional Left–Right alignments and draw in voters from both sides of that divide.

There is, however, no reason why strong, simple, single issue appeals should be the sole preserve of populist parties who are also climate change deniers. A dramatic environmental initiative that provides jobs and income plus a universal minimum income guarantee should have a general appeal, since both offer concrete immediate payments, benefits and solutions in contrast to the longer-term and somewhat nebulous benefits from limiting or excluding immigrants. It is a simpler and more immediate answer to perceived social problems. As Ernest Hemingway said, "the problem for the poor is, they ain't got no money". These measures put it directly into their pockets in a simple and understandable way. By divorcing jobs from income at the lower levels of employment, it should also alleviate social tensions there and trump the populist attack on immigration as the root of all ills. The fact that both programmes stem from a general analysis of climate change, its causes and remedies, also gives them an immediate applicability to local disasters such as floods, fires, depleted fish reserves and pollution – to which the income guarantee provides immediate personal solutions in the way of practical help. As the disastrous consequences of climate change develop, we may expect an income guarantee to strengthen its appeal in an increasingly uncertain world. At least it guarantees food next week. And environmental restoration offers some hope and confidence that something is being done to tackle future disasters, as well as providing jobs in the immediate here and now.

It is important to note that we are talking about a *guarantee* in regard to income. That is, if individual incomes fall below a decent living level, the government will top them up to that level. So it is not a universal shell out. People earning above the specified level do not get a payment. The guarantee is thus redistributive in its effects. However, those with higher incomes gain the security of the universal guarantee in a world where everything is constantly changing. In the meantime, they pay the taxes that support their own guarantee and also pay out to those who most obviously need it – unless, indeed, the government can find other resources by borrowing on its extended credit, a point also discussed below.

There are a variety of arguments that can be made against this income arrangement and that would undoubtedly be made by the other parties in the political debate. The most common is that national finances cannot afford it. Another is that those paying taxes or contributions into the scheme will resent those who have not paid and are seen as free-riding after a lifetime of non-contributing. To assess the viability of the income guarantee as a winning single issue for climate orientated parties, we have therefore to consider how the pro and con arguments will go.

5.4 Debating the income guarantee: the pros and cons

Hostile reactions to any proposal for unconditional state payments to citizens ultimately stem from the neo-liberal belief that '(s)he who does not work shall not eat' – work being defined as paid jobs. So all state benefits are to be conditional on seeking such jobs, possibly with grudging concessions for total disability, or for having made insurance payments linked to state benefits earlier. Apart from old age pensioners, persons receiving long-term state payments are generally dismissed as scroungers on 'hard working families' whose own precarious position in low paid jobs can be blamed on the jobless, as well as immigrants. Even better paid workers can be mobilised behind this argument by attributing much of their tax burden to free shell outs to the undeserving.

The greater general prosperity promised from national economic growth can also be seen as stunted by non-job-related payments. The homely argument that all can understand is that countries are the same as households, balancing annual income against expenditure and getting into serious debt problems if one does not equal the other. From this point of view, non-taxpayers receiving state payments (as well as the general public services provided for them) plunge the country into greater debt and thwart plans for future prosperity for all.

We have already discussed the fallacies of this view in Chapter 3. States are not private households. In the international financial world, they run on expectations of being able to pay reasonable interest on secure loans rather than these being tied to the one unique strategy of balancing current revenues against expenditures. Financial confidence in state ability to repay loans with reasonable interest has in fact been the basis of the national debts that have been accumulating in developed countries for hundreds of years. Pumping money into failing banks and increasing money supply to stimulate business – thus piling up more debt – strengthens general economic confidence rather than undermining it when presented as 'quantitative easing' instead of an additional payment to rich shareholders. All's in a name! Under an increasing threat from the heat explosion, spending on climate calming measures and an income guarantee to cushion the population against it is more likely to boost financial confidence that there is a coherent plan to cope with the looming crisis, while at the same time boosting economic activity.

The nebulous character of national economies and finance based on highly abstract concepts like GDP and balance of trade (often, however, leading to grim social and individual consequences) is certainly a point for campaigners to stress in political debate. It is strengthened by the unprecedented borrowings, expansion of credit and expenditures undertaken by governments as necessary measures to keep business and economic arrangements going during the Covid lockdowns. However, the argument will inevitably be made that these need to be paid for by public service and welfare cuts rather than by changing priorities and cutting the vast administrative overheads that the current welfare system involves.

Reducing 'big government' by direct individual payment of benefits is one argument that appeals to free marketeers. Another is extending freedom of choice by allowing more people to make their own decisions about what to spend their support money on, rather than being bound by the conditions attached to most welfare payments. From an extreme libertarian viewpoint, all public welfare and health services could be abolished in favour of giving all citizens a direct income payment to spend as they pleased (not an argument, however, that climate campaigners need endorse!).

This suggests that a guaranteed income payment could attract support from all sides of the current political divide. It also underlines the fact that proposals for providing a basic income come in many forms. Most are for a universal payment to all citizens of a fixed amount of money regardless of need – usually this is a relatively small amount, given that it is spread out over the whole population. The surprisingly large number of cases where such public payments have been tried out or actually implemented do show that it has significant health and psychological benefits for the poorest recipients. And by letting the better off also benefit it reduces resentment and potential political opposition.

On the other hand, universal payments divert money to those who do not need it while reducing the amount that could be paid to those who do. From the point of view of climate change, where environmental counter measures necessarily involve people losing jobs in poorer parts of countries, small payments are not enough to compensate for their loss of income. So political objections to climate calming measures will not be warded off by this measure. Only a decent living income would be enough to compensate. An income guarantee would thus work better in the climate context than a universal basic payment.

Potential resentment that non-contributors were benefitting could be reduced by topping up the guarantee proportionately, above a minimum decent living level, for those who have previously contributed through taxes, so they would get a bit more. Anticipating an increasingly uncertain world of pandemics, fires and floods, however, universal immediate security should have considerable attractions even for those currently well off but in danger of losing their job and assets overnight, or seeing their neighbourhoods devastated by fire and floods.

Potential resentments against an income proposal substituting for the incredible maze of benefits, different kinds of pensions, special grants, conditional payments, housing grants, etc., made by governments under current state welfare schemes could also be met by allowing anyone who is presently or prospectively receiving them within the next ten or 20 years to stay with what they are getting at present, if they wish to. The minimum income level is likely to be more generous than present payments so relatively few would choose this option. It does, however, provide a reassuring freedom of choice for those who value it. This would reduce the kind of opposition based on suspicions of losing out under the reforms that President Macron encountered in France when trying to consolidate the state pension system in 2018–2019.

Not only does a guarantee rather than a universal payment leave more money available for those who really need it. It also allows for the individual income guarantee (relative to local standards of living) that richer countries need to provide for the underdeveloped and developing world, as compensation for safeguarding the natural processes that absorb carbon, and for limiting emissions and waste from their industry, agriculture and fishing. We have already mentioned direct payment of income to individuals as a prime means of stopping harmful activities. We will discuss how developed democracies might provide finance for this (partly by saving on their own harmful activities themselves, such as wasteful defence projects and needless wars [Chapter 8]).

Transferring resources from the rich of rich countries to the poorer residents of poorer ones also unblocks one of the major obstacles to agreement on calming measures at world conferences on climate. That is, the demand for compensation by countries threatened by or losing out by the changes primarily driven by richer countries' activities (island countries threatened by rising sea levels for example). Direct income payments to the poorest are the most efficient way of providing such compensation and remove problems of corruption and diversion of the funds. Of course, income top ups would be made relative to local cost of living rather than all being paid at the most generous level. This would help keep payments within the limits set by world resources. Most of the world population cannot expect ever to have the level of support that democratic (mainly Western) citizens enjoy now. To provide money for a limited income guarantee (even one related to local costs of living) over the whole world, actual payments must be made directly by donors only to the needy – a point we make again when considering the international arena (Chapters 7 and 8).

5.5 How to get support to pay for an income guarantee

The clinching argument against providing universal guarantees of a minimum living income is that even rich democracies could not afford it. This has always been an argument against any expansion of welfare, from a time when it consumed only 1 percent of national GDP (as measured at the time) up to present levels. This of course may well be true if current government priorities and ways of doing things in most democracies are respected. And it is often an election-winning argument for voters who take the current arrangements for granted and concentrate on the short-term issues that have come up around election time. Expenditures of US$80 or 100 billion on high speed transport projects; enormous subsidies to large companies to keep going in order to provide employment; cripplingly expensive new weapons systems or aircraft carriers to sustain a global role against perceived threats; maintaining prisons to house increasingly large sections of the population; subsidies and infrastructure to attract new industry – all tend to be accepted as given, thus ruling out any increases in direct personal benefits as unaffordable in the face of such 'necessary' expenditures.

Some challenges to these have already been made, attracting considerable support at elections. In the early 1970s a tax lawyer, Mogens Glistrup, stood on a platform of abolishing both the welfare system and the armed forces in Denmark, substituting for these a recorded telephone message in Russian: 'We surrender'. His new party, the Progress Party, won such support in the 1972 general election that a regular government could not be formed. Over the years, however, the Progress Party was sucked into the system and became a regular member of coalitions that retained both the army and welfare – rather like the fate of the Five Star party more recently in Italy.

The enduring lesson from their initial electoral success, however, is that voters can be mobilised behind a radical agenda that challenges the existing order of things. The extraordinary measures taken by all governments in 2020 to stave off mass unemployment and economic collapse set a precedent that shows that a reordering of priorities never undertaken before can be successful (more or less) in the face of impending catastrophe. As climate-related emergencies cumulate therefore, that will be a strong basis for mobilising electoral support and a striking precedent for action. Only, it does require Green parties and ecological movements to adopt radical populist tactics and measures, and to mobilise behind charismatic leaders prepared to attack the whole of the political establishment on the lines of Grillo, Salvini, Le Pen, Farage and Trump – but with the very different message that catastrophe is just around the corner and ordinary citizens need to be protected from it.

Of course, success can never be guaranteed in politics, but carrying on with 'responsible' policies as usual has not worked for Green parties even given the urgency of the situation. The Covid pandemic has demonstrated that 'The times they are a-changin'. Convincing evidence that they are changing – politically as well as naturally – is all around, and points to the kind of evangelism that creates issues and mobilises supporters to extend the democratic boundaries of action. Violence is no answer since states are well equipped to contain it. Mass evangelism with a clear and reassuring message to support people in the face of immediate disaster is what is needed – something like Extinction Rebellion but applied to immediate guarantees and benefits, and no compromises in election campaigning.

Can we get democratic action? The jury is out. But the opportunity is there. It has to be seized if anything effective is to be done, starting with the promise of personal security provided by a universal income guarantee, and going on to a coherent environmental initiative to tackle climate change and boost economic recovery and jobs after Covid.

5.6 Further reading and reflection

Some commentators have argued for a change in voting procedures as the way to get action against climate change. It is certainly true that general elections, which try to match Parliamentary seats to party votes (various systems of Proportional Representation (PR)) help newer parties make their mark quickly rather than blocking them, as single-member constituency systems do. Afterwards, however, the politics of compromise in multi-party parliaments tend to divert Greens from concentrating on the climate cause.

A more radical proposal is to espouse direct democracy – having direct popular votes on policy (referendums and initiatives) and extended discussions in citizens' assemblies and other popular gatherings (Rebecca Willis, *Too Hot to Handle: The Democratic Challenge of Climate Change*, Bristol, Bristol University Press 2020). Local referendums have often produced votes against ecologically damaging developments. The outcomes from national referendums are more uncertain, however, and might well favour jobs over environment. So it seems a better strategy for Greens to fight directly for emissions reductions rather than for procedural and constitutional changes that might or might not produce climate calming measures in the long-term.

The most comprehensive and updated review of the whole idea of a state guaranteed basic income is provided by John Lanchester in the *London Review of Books* (18 July 2019, pp 5–8). Its introduction is made more urgent by the impact of globalisation and growing inequality within countries (see Thomas Piketty, *Capital and Ideology* [Cambridge Mass. Harvard University Press 2019]). Both globalisation and its effects, the social gains of a minimum income guarantee, and ways of paying for it are detailed within the British context in Ian Budge with Sarah Birch, *National Policy in a Global Economy: How Government can Improve Living Standards and Balance the Books* (Basingstoke UK Palgrave Macmillan 2014) Chapters 2, 4 and 5.

A supporting but independent analysis of what needs to be done in the post-Covid economy has been made by Professor R. MacDonald, a leading economic adviser to the International Monetary Fund (IMF) and the European Central Bank, in *The Big Read, The Herald*, 11 October 2020, pp 32–58. He points out that the worst response to the economic crisis would be the classic neo-liberal one of 'balancing the books' by cutting core government expenditure. Instead, money needs to be put directly into consumers' pockets since without it there is going to be no demand and therefore only limited economic activity.

Keynesian thinking of this kind, prevalent in the immediate post-war years, was discredited in the 1960s and 1970s as increased government spending failed to stimulate national economies and instead led to inflation and a collapse of financial confidence. This in turn led to higher unemployment. What was not realised at the time was that the failure of national spending to boost the local economy was due to global economic influences taking over from national ones, even within individual countries, so national governments could not stimulate domestic activity on their own. Within the global economy, however, a concerted push by national governments to spend money and create jobs by tackling climatic problems would have simultaneous effects on all countries – if only enough of them could get back to a Keynesian approach on a world level. The victory of Biden and the Democratic party in the US federal election of 2020 may trigger this, but it needs to be taken up by climate campaigners in the ways outlined in this chapter, if it is to be consolidated and pushed further.

6 Organising to expand democratic action

6.1 Mass organisation and mobilisation

The last chapter focused on expanding mass support for ecological action through political parties pursuing relatively conventional democratic tactics – arguing, debating, competing in general elections, winning elective office and promoting popular referendums and initiatives. The key to success in all of these is having policies to alleviate the human losses involved both in climate change and the measures necessary to avert it. Such measures would provide an economic boost and jobs after Covid. But to tackle growing social and political disruption, climate campaigners also need to offer the guarantee of an income sufficient to maintain a decent living standard for every individual citizen, geared to each democracy's cost of living and with social safeguards for those with mental and other problems that would affect their spending. The side benefits of the income guarantee in terms of freedom of choice, reduction of administrative costs and interference and the various approaches to financing it have already been discussed, so they need not be elaborated here.

What does need to be emphasised, however, is the organisational support Green movements need in order to bring this policy to the forefront of political debate and make it the specific issue that dominates elections. As this book has emphasised from the Preface onwards, no political cause can succeed if not does not innovate and organise – also in terms of democratic action. Individual commitment and action are essential, but they will ultimately be futile against the powerful, established, wealthy forces confronting it – especially when these effectively control the state apparatus and most of the media. These have inhibited and constrained effective climate action so far. This is partly because campaigners have followed conventional routes in promoting their policies – public debate and discussion, political party activity

DOI: 10.4324/9781003221630-6

and street demonstrations, which have sometimes, but not usually, descended into violence.

The democratic establishment of existing parties and their governments, acting through a supporting state apparatus, have been well able to contain such movements. Electoral hurdles have left Green parties small. It is not that governments have done nothing in response to such pressures – Chapter 1 has listed the climate calming developments under way. These are not insignificant, but certainly not enough to meet the mounting crisis of the 2020s. And palliative measures are themselves used as a tactic to head off more far-reaching Green and ecological demands. There are always other short-term issues and concerns that take priority in the governmental and party agendas, and divert popular attention away from climate.

The limits to current action are well illustrated by the Extinction Rebellion of 2019. Innovative in its international reach and decentralised organisation, it focused on headline grabbing actions, intervening at the UN and international meetings, most successfully shutting down the centre of London and other cities for a week in the summer of 2019. It certainly prompted debate. But where is it now? Brexit and pandemics grabbed attention and took over the headlines and the crucial climate conference of 2020 was postponed by the UK, the very country where most summer actions took place.

Certainly, mass demonstrations and electoral and legislative pressures should be continued. To be more effective, however, they need much more coordination and continuing organisation than they have had, especially if they are to sustain the positive changes in lifestyle that have come with the Covid pandemic and the likely pandemics to follow. Extinction Rebellion – or some organisation like it, with an evangelical fervour and popular appeal – needs to have continuing marches and meetings organised by branches in every city and district, something like the weekly 'marches under one banner' for Scottish independence, or the Gilets Jaunes in France (without their violence). Lessons can be taken from the evangelical and feminist movements of the late 19th century, or Methodism earlier on, with individual pledges collected to take action of various kinds. These could cover individual abstinence from emissions to changes in lifestyle and consumption, boycotts of plastics and meat and participation in mass marches and movements. Local chapters should operate with programmes for social and ecological activity. Walking and picnicking, bicycling, climbing and canoeing, cultural and countryside excursions – all are hugely popular activities that could be given an ecological flavour by Green parties and climate activists. Membership drives could be launched among the volunteers in conservation groups such as National Trusts, walking organisations, social clubs in flood-prone or fire-prone areas or local groups opposed to environmental destruction.

Cultivating and organising the social sphere and then drawing on the participants to promote their overall political cause in various ways was indeed the way that political parties of all persuasions operated in the early 20th century, so creating a popular base to build themselves up on. Nowadays such direct organisational activities can be reinforced by continuous use of social media to create networks and news flows supportive of climate gatherings and causes, as well as linked leisure activities. Dedicated TV and radio stations can also be created for this purpose, with scientific, cultural and political content, both educational and entertaining, to further climate action. Reaching out for mass support – always the goal to be borne in mind – will rest heavily on the efforts and money of disproportionately middle class and professional groups, as has always been the case with political movements. Social action groups, with their emphasis on a minimum income guarantee as an individual safeguard against current disasters such as the pandemics, fires and flooding now emerging from globalisation, could, with their impact on the worse off, mobilise the masses as unions, cooperatives and social clubs have in the past. A useful start would be practical instructions on how to get groups going and organise neighbourhoods, instead of operating purely on an intellectual and educational level. Another would be collecting money for paid organisers to direct and support these efforts.

Whereas in the past Greens and ecologists have relied primarily on reasoned argument and conviction – easily brushed aside by more imme-diate and 'practical' concerns – the focus now has to be on continuing financial and organisational effort to bring immediately appealing concerns to the fore, channelling the passions and charisma of Extinction Rebellion into continuing organisation and institution building. We have had the outbursts of passion, which need to continue, but now they also need channelling into systematic efforts and permanent processes. Kickstarting this process requires money. There are enough charitable foundations and concerned individuals to provide this once a coherent, overall action programme is formulated. And there is enough talent to formulate that, once the need is realised. But it must be done now.

6.2 Innovative political action

Even keeping up current efforts, and strengthening and expanding them by conventional party-political means, does not make it certain that activists can galvanise democratic governments into necessary climate action before it is too late. Recurrent pandemics with their quarantines and social distancing will cut down the practical opportunities for meeting and organising. Normal democratic processes might get us round to accepting and implementing new, drastic, radical policies in the end, but this takes time, which we do not have. Moreover, normal processes and compromises tend to water policies down until they become palliative rather than really remedial. We have seen this with the very limited political success of Green parties over the last 40 years. Desperately needed large scale changes have simply been delayed or side-lined, to the point where we stand on the brink of total climate disaster. It is true that electorates have become more volatile and for various reasons suddenly switched from established to populist parties. While this favoured the broadly ecological Five Star party in Italy, it also boosted climate change deniers there and – famously – Trump in the United States. Current electoral politics thus offer chinks to climate change campaigners to influence policy, but they also offer manifold opportunities for climate deniers to divert or block it and to discount the possibility of any catastrophe at all.

What is to be done then to supplement the prospects of success from conventional democratic action? What we suggest here is taking unconventional democratic action – unconventional, that is, in the sense of not being generally utilised by climate campaigners or other progressives, though by plenty of others, and democratic in the sense of being based on policies adopted by democratic governments themselves. These strategies do not go against national or international law, but they do hit at the basis of state power as analysed in Chapter 2. They create a fiscal crisis of the state that can only be met, within the time available, by capitulating to demands for the radical actions necessary to reverse climate reversal.

These strategies also have the advantage of drawing on the basic strength of the climate change movement in terms of its core professional, highly educated and relatively wealthy base. Getting mass support outside this social grouping is really important. However, it is also problematic. Democratic electors and governments are quite likely to take drastic action once the climate crisis becomes clearly catastrophic. This has been dramatically illustrated all over the world by the Covid crisis. Unprecedented quarantines and unlimited spending were precipitated by threats to the national health and economy, but very little was done beforehand to avert it. Health and social services were cut down, and a mushroom (gig) economy was promoted that collapsed immediately quarantined consumers stopped buying. Voters reinstated governments committed to growth at all costs, with no regard for public services and little for measures against climate change, right up to the point at which the pandemic arrived – and may well continue after it is seen off.

The only dependable political base for climate movements therefore are its convinced adherents, mostly professionals converted to the cause by the developing climate sciences. Although a minority within democratic populations, this will be a growing base and one with substantial resources that now need to be utilised for the climate movement. In particular, since both businesses and the very rich avoid paying taxes to national governments through various (perfectly legal) strategies, the middle stratum of society from which most climate campaigners are drawn increasingly forms the main set of taxpayers in democratic states. The less well off, though often taxed more heavily in proportion to income, cannot in aggregate match their contribution. And, of course, many at the lower economic levels perforce draw income from the state rather than contribute to it. Their voting power may be crucially important in influencing democratic elections and policies, but it is of less importance in putting on financial pressure.

How is this to be done? As previously suggested for the electoral sphere, the answer lies in innovation and organisation. Individuals acting alone or in non-expert groups will undoubtedly find it difficult to withhold tax payments for government policies they disapprove of, and divert them to those they do. They will also find it difficult to locate and pay tax advisers who can help them do this. Large ecological institutions and advisory services can, however, employ experts to ease or even perform such services for individual supporters. If enough then succeed in channelling potential tax money elsewhere, the pressures on democratic governments to change policies in their preferred direction will be immense. But there must be organisation and the will to generate it. A first obstacle to overcome is psychological. In advocating tax avoidance on a massive scale to push governments into climate action, we have to convince folk who are, for the most part, passionately democratic in outlook, who actively support good causes and public services, and who are committed to moral standards and peaceful persuasion, to take the kind of action usually favoured by shady businessmen.

Climate campaigners, on the other hand, would most likely prefer to take moral stands, ending up in court through dramatically refusing to pay tax or for mass obstruction and civil disobedience. To get the support of such people, they have to be convinced that their taxes can be diverted to ecological and climate calming ends, even if handled in low tax countries under opaque procedures that they oppose on principle.

Another problem in convincing these admirable people to channel taxes away from their government is their general support for public services and greater government intervention (necessary to combat climate change anyway). Eroding the tax base also gets in the way of implementing the minimum income guarantee. Governments are likely to be cutting public financing anyway in the wake of the extraordinary subsidies offered to businesses and employees during the Corona pandemic to ward off economic collapse and general destitution. A tax rebellion could accentuate this tendency, resulting in social hardship and even a turning away from climate calming measures as unnecessary impediments to economic recovery. Governments also have their own priorities and control over the distribution of tax revenues, so money you would willingly pay for public services or propping up incomes is often spent by governments on other ends, such as expensive defence, construction contracts or payments to banks ('quantitative easing'), which ecologically minded tax payers would like to see terminated.

The main argument that would help win over these campaigners to tax avoidance schemes within their own state is that, channelled properly through an ethically minded foundation operating within a low tax jurisdiction, the money that would otherwise go into general taxation for spending as the government wants could be given directly as charitable donations to support the services the taxpayer approves of. Distribution could be organised to best effect by charities located in the home country who could distribute the money either to public institutions such as health or welfare services or to other charities working in the field. The details need not be discussed here, but the principle that the tax payments avoided at home would ultimately go into causes the taxpayer approves of should be reassuring. Among these, of course, could be minimum income support and climate calming measures. The intention is not to divert money for personal aggrandisement, as multinationals and millionaires benefitting from tax avoidance do now. It is rather to exert ethical control of the money paid at the national tax level so that it goes on approved policies and ultimately pressures governments into implementing them – whereupon tax money can stop being diverted and be paid domestically as normal. As we have stressed, all this is perfectly legal and indeed promoted under the legislation passed and upheld under democratic processes in most countries.

The details as to how medium salary earners might be mobilised and helped to do all this are filled out in the next section. At this point we simply need to stress what a potent political weapon ethical tax avoidance could be in getting democracies to take appropriate climate action both at home and abroad. As pointed out in Chapter 2, all state action in providing public goods and pursuing the policies necessary to produce them rests on their ability to extract the necessary resources, mostly from their own population and territory, and by credit-worthy borrowing. What a threat, therefore, is any weakening of their ability to do so by reducing the revenues they get to spend as they wish!

In politics threats to take some line of action, if they are convincing, are often as important in shaping events as taking the action itself. This is particularly true where financial confidence in government credit-worthiness is concerned. Anticipating major trouble, democratic governments often change policies in order to boost confidence. However, the threat has to be credible in order to be convincing, so climate campaigners would have to take realistic organisational action before publicising what they intended to do. These would include hiring skilled tax advisors and accountants and surveying low tax countries that would accommodate the firms, businesses and trusts through which revenues would be transmitted. At home initial funds would have to be raised from supporters to do this, and individual pledges collected to follow through on these procedures to avoid domestic taxes, divert the money abroad and then to channel it back as charitable donations to approved public services (and/or to international climate calming activities). All this organising will take time and can hardly be kept under cover, so one possible reaction by governments, instead of prioritising and extending their climate action, could well be amending legislation to close the tax avoidance loopholes being used. Their problem, however, would be that these are precisely the same loopholes used by domestic businesses, multinational corporations and the very rich to save on their taxes. And changing them (often used as an excuse for delaying and avoiding any reform of taxation at all) requires either international agreement or drastic unilateral action with unknown economic and international consequences. International negotiations are proceeding on this at the present time, but outcomes are still in doubt and depend on the Democratic party staying in power in the US.

Even the European Union, which has actually tried to extract back taxes from the multinational Facebook, has been ruled out of order by its own courts, and the UK has never followed up its vague talk about the possibility of realistic taxation of the multinational's British turnover. Therefore, the only way that most governments could thwart climate related action in this area would be by forbidding charities to use the tax avoidance schemes available but exempting businesses and the rich! Drastically unmasking their opposition to effective climate reform measures would, however, be politically difficult in a period when climate threats are growing and lip service is increasingly paid to the goal of reducing emissions. Much better to quietly downgrade effective climate action in favour of economic growth, while still offering vague general support to it – the situation ethical tax avoidance aims to change.

Organised, ethical tax avoidance is a win-win-win-win strategy for climate campaigners. First, it draws on their major basis of support in the well-educated professional classes, earning more than the average and the major source of tax revenues. Second, it hits at the major basis of state power – the ability to extract the resources they need to pursue their preferred policies, which are more likely to be preparations for war and stimulation of economic growth with increased emissions, than a minimum income guarantee or stabilising world climate. Third, the action, though unconventional, is entirely within the legal bounds set, and indeed promoted, by democratic governments themselves. If it then prompts them to change their minds about allowing tax avoidance by businesses and the rich on a massive scale, that would be a good outcome in itself, since it would solve the fiscal crisis of the state that has so often been an excuse for avoiding climate action. Fourth, ethical tax avoidance can channel tax money by direct payments back into the domestic policies campaigners approve of, such as health and welfare. It can also voluntarily provide money for a minimum income guarantee, avoiding accusations of inconsistency in what climate campaigners are advocating. So what's not to like in general about this strategy? Now we turn to a more detailed account of how it can be organised and carried through.

6.3 Organising an ethical tax revolt

Obviously a first requirement for organising any political action is to obtain the agreement of supporters and sympathisers to it. In this case one needs to go further by collecting individual pledges to take part in the action once it starts, and to contribute money to setting up an initial organisation to handle technical details. For the reasons outlined previously, climate campaigners are likely to be viscerally opposed to any organised schemes of tax avoidance, preferring individual stands and overt confrontations to the perfectly legal scheme proposed. Individuals can exert a powerful influence by martyring themselves – often in demonstrations that go beyond the legal limits. However, for over 40 years their efforts have not produced the changes needed. This is a powerful argument for convincing supporters that an entirely legal collective action can be more effective, if coordinated on a mass, organised and continuing basis. Assuming that the necessary accountants, tax advisors and lawyers can be hired, the next step would be to produce a plan of action for diverting individual taxes. This would have to differ between countries as the relevant legislation itself would differ. At a general level, however, it would involve:

a   Identifying the most suitable tax havens, usually mini-states and often British dependencies, from the Isle of Man and the Channel Islands to the Caribbean.

b   Setting up shell businesses, funds and charities to handle the money coming in and re-channel it to each individual's approved charities, causes and public services in their home countries and/or to international climate and ecological action.

c   With institutional machinery in place, individual pledges should be activated. What individuals need to do in their usual situation as employees or pensioners is to opt out of situations where their tax is automatically deducted by their employer or pension fund and paid directly to the government. They can do so by either requesting directly to change arrangements so they handle their own tax payments – as in the case of their private investments, for example, they already do. Or they can set themselves up as self-employed or singly or collectively as small businesses, negotiating a contract that gives them the same rights as they had as an employee but that allows them to deal directly with their own taxes. Easier said than done, and probably too arduous to do on your own!

d   However, this is the point of having an organisation to make these arrangements for individuals who would otherwise find the details impossibly technical or time consuming.

e   Alternatively, individuals could transform themselves into employees or shareholders (or both) of a shell company headquartered in a tax haven, getting their normal income paid back to them while the difference between domestic tax and re-domiciled tax would be paid out (as under b) above) according to their specifications.

Given the different tax regimes and laws in different countries, it would be unwise to go into greater detail on the tax avoidance strategies to be used. That, after all, is the reason for hiring experts and advisors in the first place. Something along the lines sketched above must be feasible, however, since so much tax avoidance already goes on to make the rich richer and multinationals grow. But why should Mammon have the best schemes? The important thing is to innovate and organise, and the first step in organising is to get expert advice.

In general, unless campaigners exert pressure on governments where it really counts for them, they will get no more action by governments than they already have – and the climatic tipping point will be reached by the mid-2020s. The imperative is to act practically and effectively now before irreversible climate changes are set in motion. And action must be collective and organised broadly along the lines sketched in this and the preceding chapter if it is to be effective at all.

6.4 Further reading and reflection

Detailed analyses of accounting practices that facilitate evasion by businesses, multinationals and the rich have been made by Prem Sikka, Professor of Accounting at the University of Essex:

Sikka, P. (2010). Smoke and mirrors: Corporate social responsibility and tax avoidance. Accounting Forum. 34 (3–4), 153–168
Sikka, P. (2011). Accounting for human rights: The challenge of globalization and foreign investment agreements. Critical Perspectives on Accounting. 22 (8), 811–827
These were further elaborated in a symposium in *The Guardian* (www.theguardian.com) in the week following 13 December 2014.
One aim of ethical tax avoidance would be to get national governments to abolish the loopholes that make it possible (and that allow multinationals and the rich to benefit themselves). Were the policies it is designed to reinforce then adopted by national governments – particularly the minimum income guarantee and climate calming measures – ethical tax avoiders would then of course start paying all their home taxes again in order to support them. Ethical tax avoidance is not a principle of ecological thinking but a potent, if temporary, political tactic to be used to get necessary government action.

7 Climate action in non-democracies

7.1 Introduction

Collective, organised action to combat climate change is easier to undertake within a democratic setup, where governments are supposed to respond to popular opinion, than under other types of regimes where they strive to control it. Since the Soviet Union, with its controlling Communist Party, made its extraordinary transformation into present day Russia in 1990, the prime example of a single party based on an authoritarian top-down regime has been China. But there are plenty of other governments, around one third of world states, that suspend or manipulate procedures to keep themselves in more or less exclusive control of policy-making and implementation.

Typical means of doing so are censorship of opinion in the social media as well as in the press, broadcasting and publishing; imprisonment, torture and murder of political opponents and sometimes anyone critical of the regime; and strict control or outright bans on any organisation not sponsored by the state, including international or foreign ones operating within the country. In extreme cases minorities and religious groups may be 're-educated' or dispossessed, massacred or forcibly assimilated. While maintaining a facade of elections and parliaments to bolster their claims to be 'with the people', these are manipulated in various ways to ensure that they always produce results acceptable to the current government.

These features, of course, occur in less overt forms in many democracies. Political assassinations, suppression of opposition, imprisonment on trumped up charges and interference with press freedoms occur in various ways, even if relatively free elections and governmental change still occur under these unpromising conditions. There is thus no hard and fast distinction between democracies and non-democracies. Procedures and behaviours are always subject to evolution and change.

DOI: 10.4324/9781003221630-7

There is a continuum going from fully-fledged stable democracies with full individual rights, and freedom to take initiatives like those discussed in the two preceding chapters, through those where you can do so but at some peril, on to those where free initiatives are totally ruled out and automatically result in torture, imprisonment and/or death.

What then can be done domestically to promote climate action under these conditions? Where regimes are less repressive and organised, there is, of course, always a possibility of pushing at the barriers, but under more efficient repression, popular initiatives are probably doomed to failure. What domestic strategies and tactics can be used instead? We concentrate on domestic developments here, although one answer – possibly the most potent one – is to get democratic governments that do respond to the kinds of popular pressures discussed in Chapters 5 and 6 to exert external pressures on the other countries at an international level. Ways of doing so are mostly considered in the next chapter, but steadily increasing globalisation means that internal developments inside one country are always going to be influenced by what is going on elsewhere.

Globalisation ensures that there are always communities and groupings inside each country with close links outside it, and which are influential because they are useful or indispensable to the regime (professionals, scientists, doctors, engineers to name a few) or because they are too well known within the country to eliminate (usually religious groupings).

The leadership and members in both cases may be working with the government, but their skills and status give them an independent standing. So if world opinion shifts within their profession or grouping, most freely expressed in democracies, they are likely to follow it to some extent, even if it goes against official tenets in their own country. Since scientists and related professionals are the most obvious carriers of opinion here, we begin by considering their role in promoting climate-related causes under non-democratic regimes, focussing particularly on China since it is the largest and most repressive example. However, if progress can be reported from there it is also possible under less extreme conditions elsewhere.

7.2 Scientific and professional promotion of climate-related measures in non-democratic settings

Most authoritarian regimes run poor countries with social problems to which economic growth and development seem the only solution. This is true for many democracies too. The difference is that these democracies do not necessarily view foreign interventions in the ecological and climate areas as a threat. They are thus open to actions such as foreign buying up of forests and other carbon reducing terrain, and potentially to the most general and effective action of all – allowing an internationally administered guarantee of a minimum living income to make up for stopping destructive environmental action.

Under a military or other coercive regime, on the other hand, foreign interventions of this sort are viewed as threats to their power and opposed for that reason. All domestic initiatives are likely to be viewed in the same way. Even dictatorial regimes, however, and particularly those with a developmentalist ideology focused on economic growth, need experts, often trained abroad, to produce the infrastructure for development, such as roads, airports and dams. These, of course, are hardly climate-enhancing initiatives and are most likely to be aimed at facilitating logging and extraction industries that promote climate change rather than retard it. The engineers and technologists carrying through such developments are likely to focus on them in isolation rather than seeing the broader consequences of what they are doing. A pertinent example is damming the Blue Nile in Ethiopia, which threatens the annual fertilisation of Sudan and Egypt downriver, with climatic and ecological changes thrown in.

One immediate step in this regard is to build climate science, with its ecological consequences, into engineering syllabuses throughout the world, in order to equip engineers and technologists with the ability to give expert advice on whether construction will actually achieve its intended goals. There are many examples of elaborate constructions that have not achieved intended effects or produced unintended consequences, thus vitiating the wider developmental goals they were intended to achieve. By making ecological and local or regional climate change assessments a necessary element in construction proposals and a requirement of professional practice wherever undertaken, the technical advice on which even authoritarian governments have to rely could be a vehicle for climate calming action. Reinforcing this, all professionals could be required to take the equivalent of the medical Hippocratic oath with regard to environment and climate change – to always put these considerations first in considering any developmental project. International professional bodies might enforce these requirements and consider complaints, as medical associations do for doctors today. These would not infringe on any one country's autonomy. And even authoritarian regimes might baulk at employing unqualified or disqualified engineers or technicians, or even doctors under a broader Hippocratic oath, since climate change is increasingly seen as affecting health. In this way educational institutions and curricula, with the most prestigious and advanced professional training done in democracies, can influence climate policy throughout the world, even under authoritarian regimes, without requiring democratic governments to act as intermediaries.

Such action also depends, of course, on the further and rapid development of the climate sciences themselves in establishing climate change as the ultimate influence on all human activity, and therefore requiring due consideration in this context of any new scheme before any actual work is undertaken. (The Paris Agreement of 2015 went some way towards this.) One can hardly say that climate scientists lag behind in developing this position. Indeed, the expansion of the subject and the steadily cumulating knowledge of how climate works, are undoubtedly the major achievements of the whole campaign against change so far. It has altered public and political opinion, although not fast enough. It is the major driver of all government advice in terms of orchestrating measures against climate change at both national and international levels. It must, however, insert itself into every degree course and every university and institute at the scientific level if it is to influence developments in every part of the world, as it needs to do. (An initiative going in this direction has been undertaken by Scientist Rebellion. See https://scientistrebellion.com.)

The major weakness of climate science here is its inability to predict authoritatively when and where irreversible change will actually take place. As we have noted, the credibility this lends to an incremental interpretation – natural processes of carbon absorption remaining more or less constant while human emissions gradually decrease – gives governments and their agents leeway to postpone any real action while they get on with their immediate concerns of active and cold wars, massive climate changing construction, logging, mining and fishing and national economic growth at all costs. These mesh in nicely with an election cycle of 4–8 years. Climate problems can be left to the next government, which in turn leaves any real action to its successor.

Of course, the two theories of climate change pointing on the one hand to reasonable natural stability if human emissions are progressively reduced, and on the other to a looming cataclysmic reversal very soon, both agree on what needs to be done – reducing emissions as quickly as possible. It is the degree of urgency and the exact balance of climate-reducing to other economic, social and military considerations that is in question between them. To be generally convincing, one theory needs to be tested against the other in the classic scientific way, by formulating each clearly enough to make precise predictions of the immediate course of events and deciding between them on the basis of which makes the most successful predictions of what will happen in the near future. Of course, work in this direction has been proceeding for many years. Since it is now so immediately relevant for human – not to say planetary – development and survival, we do need to have an authoritative verdict now, perhaps from a world scientific congress independent of governments with relevant predictive evidence before it.

The more theorising and evidence can predict what will happen in specific countries and regions, the more likely it is to influence action in them. The most obvious example is China. At once the driver of emission increases elsewhere through its ever-expanding demands for raw materials and food, it suffers itself from enormous ecological problems from emissions, threatening health, and on the way to rendering cities uninhabitable; environmental degradation, a danger to long-term food production; and disastrous flooding and droughts from massive dam building and river diversion. All of these would be brought to a head by climate change. If cataclysmic climate change was authoritatively verified and specific disasters inside the country could be predicted for dates in the 2020s, the process would become an immediate rather than a postponable problem for the government. Additionally, it would add enormous clout to their scientific experts within the governing hierarchy. Were climate scientists united round a validated theory of imminent catastrophic change and able on this basis to predict regional effects – and particularly if the subject was a compulsory element of every scientific and technological syllabus throughout the world – one could well imagine scientific advisers exercising a decisive influence on decision-making even in such an authoritarian system as the Chinese.

Under more open systems where effects could be more widely debated, prospects of imminent disaster would also concentrate minds, particularly if specific regional and country effects could be spelled out. Of course, more open debate would provide space for more opposition. Democratic decision-making is, by its nature, slower than where policy is imposed from the top. However, envisaging a situation where climate science has established itself as authoritative and reasonably specific predictions can be made for each country or continental region, one can well expect a stronger political consensus emerging for accelerated action, as in the recent case of the Covid pandemic.

Hopefully the force of the scientific agreement would prompt preemptive measures rather than the panicked reactions we saw arriving too late during that crisis. A few countries, mainly in East Asia, and both authoritarian and democratic, had learned from previous pandemics and prepared adequate counter measures as advised by virologists. Few others, with the notable exception of Germany, had made any preparations at all even in terms of stockpiling equipment or providing hospital space, even as the virus was spreading round the world. This was in the face of a united scientific opinion spearheaded by perhaps the most authoritative international authority of all, the World Health Organisation. There was a lack of any cross-national coordination, as each country made its own decisions about what action to take. (In the US and Brazil, it was each state within the Union, paralleling the situation within the EU, where each country initially took its own measures in complete independence of the European Commission.) Some governments tried to downplay the existence or seriousness of the pandemic. Again, there are close parallels with climate change!

In the case of recurring pandemics, governments can, of course, learn from experience and follow the scientific advice before rather than after the virus has spread – also by strengthening the international institutions to deal with it. These may enable them to break the impact of the next global outbreak. One general consequence may be to give relevant experts, and scientific advice in general, a more authoritative role in decision-making even outside the medical area. This might be true also for climate science, particularly if it comes to an agreement on the speed with which changes are taking place. The problem here, of course, is that one cannot simply learn from experience as in the case of recurring pandemics, each of which increases our knowledge of how to handle and perhaps avert the next one. When climate change occurs it will be irreversible, perhaps for the next 10,000 years. Once natural processes reverse, they usher in a new geological era. The time to avert this can only be now, before the tipping point is reached. Panic measures afterwards will be too late. This makes it imperative that scientists reach an authoritative consensus very rapidly.

The obvious way to do so is to hold a world conference not dependent on governments, and scientific not political in character. Its aim should be to endorse an authoritative theory of what is happening to climate now, with specific short-term predictions about developments during the 2020s in the major countries and regions of the world. Specific short-term predictions are necessary because short-term problems are what governments respond to. Otherwise they can be kicked into the long grass for action later, preferably by the next government. Such a focused scientific congress could supplement COP26, the world climate conference scheduled for the end of 2020 but postponed for a year by the host government (the UK) in the face of its short-term concerns with the current pandemic, exit from the EU and subsequent economic slowdown. COP is regarded as a last chance to deal effectively with climate change.

However, an authoritative, purely scientific world congress, financed possibly by the UN or an international foundation or multinational, non-polluting companies, could do more for climate calming if it came up with a unified, validated theory and specific short-term micro predictions. If the consensus was for a heat explosion in the 2020s, as the current evidence indicates, this must surely galvanise governments, authoritarian as well as democratic, into immediate action even more than likely pandemics. The enormous impact that world scientific opinion can have, even in a closed decision-making system like the Chinese, can be illustrated from the regime's change of stance from the 1990s, when its computer hacking of private scientific exchanges in the West disrupted an early attempt to get an accord on world action, to the colossal project to develop solar powered electricity generation, now producing a quarter to a half of total supply. Together with its earlier one child policy, this puts the Chinese regime, once convinced of the climate threat, at the forefront of governmental attempts to reduce world emissions.

As noted above, however, this still leaves a half of Chinese electricity being produced by highly polluting coal powered generation. The powerful interests behind this are entrenched within the closed regime and vigorously defend their lucrative share of production. Within this overall situation, it is vital that the climate scientists become as authoritative as virologists have become as a result of the pandemic (when, however, initial medical warnings were suppressed by the Chinese authorities as alarmist). Having world expert opinion behind them is vital in getting their advice heeded – before rather than after the catastrophe unfolds! The most promising way to do so is to campaign for a World Scientific Conference focused on providing an authoritative theory and diagnosis of the climate changes now accelerating towards an irreversible tipping point.

7.3 Working through religious traditions

The second widespread influence on world opinion remains the major religious traditions that unite as well as divide it. This is particularly true of the two most widespread religions of the Book: Christianity and Islam, with followers in most countries of the world. Other traditions such as Judaism, Buddhism, Hinduism and its offshoots, Confucianism, Taoism and Shintoism either have markedly fewer adherents or are more confined geographically. The other distinguishing features of Christianity and Islam are having within them a centralised authority that lays down doctrine for hundreds of millions of followers world-wide. This is, of course, truer in the case of Christianity with the strict hierarchy of the Catholic Church centralised in Rome and its supreme head, the Pope, but with churches, parishes and bishoprics throughout the world. Islam has a more decentralised structure but equal unity of purpose, with its holy book, the Qur'an, specifying proper practice in belief and living even more rigorously than the doctrines of the Catholic church.

Both Christianity and Islam have also deep internal divisions, with Catholicism facing challenges not only from the Orthodox churches of Russia and Eastern Europe, but also from Protestant churches, particularly in their Evangelical form, in the Americas and Africa. Orthodox Sunni Islam also extends into radical wings promoting guerrilla warfare and terrorism in Europe, West Asia and beyond, as well as facing increasingly entrenched Shia opposition elsewhere.

Religious traditions come in because they provide a background, often an unconscious one, for thinking about nature and hence climate. This ranges from the simplistic view that God intervenes actively in the world using climate events and natural disasters like pandemics to punish deviations from His will, to highly abstract philosophical treatises about humanity's proper behaviour within God's creation. These might take the general view that humans must act in harmony with a social and political order of which they are only part, to the opposite position that nature was provided for humans to exploit for their own benefit. The Paradise myth common to many religious traditions is ambiguous on this point. Adam names and dominates plants and animals, but too much intervention in the Garden brings about his expulsion from Paradise. Submission to the will of God in Islam may involve respect for His natural gifts or a pressing need to promote a godly earthly Caliphate regardless of other considerations.

Developing climate-friendly arguments theologically could form the basis for a mass mobilisation of believers. This might be an argument for climate campaigners to fund and support theological institutes within each of the religious traditions, which would both research and teach in this area. Inside the Catholic church the tradition of St. Francis, conversing and speaking with animals – well represented by the current Pope's adoption of his name and much of his outlook – provides an existing basis to build on.

In the related field of social justice, the actual practice of the church has often, historically, supported oppressive regimes – for example 'clerical fascism' in interwar Iberia and central Europe, now lingering on in Poland and Hungary. More pertinent to present day developments was its general support of highly oppressive hierarchies and military regimes in Latin America, with exploitative attitudes to natural resources and indigenous peoples and active opposition to social and ecological reform. In both cases, however, a major change in attitudes has developed over the post-war years. Christian Democratic parties founded after the Second World War with generally free enterprise orientations have increasingly embraced the doctrines of social and family support enunciated in the (Papal) Social Encyclicals of the late 19th century, to the point of being increasingly in coalition with Social Democrats in Europe. In Latin America and other developing areas, the development of 'Liberation Theology' has shifted attitudes from support of an oppressive status quo to active attempts to change it in a more progressive direction. While mainly focused on social justice, this movement also involves opposition to the multinational corporations whose interests are in exploiting – and in the process devastating – natural resources (and indigenous peoples) wherever they are found. Pope Francis may be seen as a product of this theological development. Of course, there are tensions and divisions within the Church itself, with most of Francis' reforms blocked or opposed at this moment. All the more need, therefore, to develop theological support for his reform project to provide a firmer base against climate change.

There is also opposition to such protective measures outside as well as inside the Catholic church. The sharpest opposition these days comes from the evangelical sects making enormous headway among the poor and dispossessed of the Americas both north and south, and in Africa. As noted previously, the poor have little choice under present dispensations but to degrade the environment in order to survive. But this is reinforced by religious beliefs that see climate change as God's will – or perhaps His punishment – that has to be accepted. To combat such attitudes, one needs a theological initiative to move Evangelical attitudes from an exclusive focus on personal salvation to one of collective redemption and harmony with nature.

Islam, of course, also has its internal divisions, most notably between the strict Wahhabi form of Sunni Islam, with its comprehensive and puritanical rules on individual and social behaviour, and Shia doctrine as enunciated from Iran and Iraq. The Qur'an does stress the maintenance of a due balance between humans and nature, but the modern emphasis in both has been on running a godly state enforcing proper behaviour on its citizens, thus ensuring their ultimate salvation. As with the Christian evangelicals, environmental degradation and climate change are minor concerns here, while the wars and terrorism that doctrinal differences provoke tip the balance even further towards ecological destruction.

In the particular case of India, Hinduism and its offshoots display more of a doctrinal affinity with preserving the natural environment. Many of its gods are nature deities or associated with particular natural features and localities (which could also be said of many local saints of the Catholic and Orthodox churches). In India this has had a particular effect in preserving many local woods and groves, also as sources of traditional medicines and material used in sacred rituals. It cannot be said, however, that Hinduism has operated very effectively at the national level to slow development and reduce emissions.

Perhaps the most important impact of all religious traditions has been on family reproduction and its effects on population growth. Most of those we have discussed indeed traditionally favour reproduction and consequent population increase. Related to this is their attitude to women. Up until recently, few have favoured empowerment of women. Respect has been given to them instead as wives and mothers within the family, usually bound up with having lots of sons. There has been positive discouragement of girls pursuing careers and jobs or having an independent income. Attitudes are changing, but again, they need to be bolstered by new theological interpretations of the founding doctrines. Particularly crucial is the question of birth control that most of the main religious bodies of all faiths oppose, apart from the established Protestant churches. It is significant that the greatest progress in this direction has been achieved by secular regimes in Russia, China and Japan, but often by encouraging or enforcing abortions. Prior birth control through women's free choices is a more appealing option.

Women's empowerment is the major factor here. With increasing freedom and prosperity, most mainstream religious groupings find that their adherents increasingly limit their families anyway. (As a partial result, world population is expected to fall by 2100.) Spontaneous birth control puts practical pressure on theologies to catch up. It would be a potent development in the climate calming campaign to develop a consistent doctrine in this area. By eliminating conflicts of conscience this could ease population pressures even more, particularly in the most ecologically sensitive (and therefore underdeveloped) areas of the world.

There are therefore possibilities for effective climate action even under non-democratic governments, whether run by a single party like China or Cuba, by the army like Pakistan or Thailand or by a theocratic elite as in Saudi Arabia and Iran. However, they do not provide the space that democracies do to promote climate friendly policies by organising to put pressure on governments. Doing so will be seen as challenging the regime itself and invite repressive action, particularly if the promoters have foreign links. This leaves international pressures – mostly from democracies – as probably the most effective way to push authoritarian governments into taking appropriate actions (and of course some democratic governments too). We deal with these in the next chapter.

7.4 Further reading and reflection

Consideration of religious traditions and their influence on attitudes to climate related processes and the general environment leads on to the question of social norms and their wider and more pervasive (if less institutionalised) influence. In many societies, polluting activities are a sign of personal success and high status. Owning a large car, going on expensive flights and ocean cruises, and wearing expensive clothes all establish your position. If these were seen as socially unacceptable, that might go a long way to curbing the consumer excesses that in turn drive environmental destruction. Governments could do something here by limiting or banning extravagant advertising. Influencing the fashion industry would be another potent way of curbing the production of cheap throwaway clothing. One success story here was the discrediting of furs – the height of fashion in the immediate post-war period – which have now become social anathema due to animal activists' agitation and publicity.

Since it goes against 'scientific socialism', the Chinese Communist regime could possibly be induced to ban traditional medicine, responsible for bringing many animals to extinction – as they have already banned live meat markets and restricted trade in ivory.

The enormous expansion of ecological and environmental awareness in the course of the last 40–50 years, spearheaded by books and media coverage, has clearly had an effect on social norms and behaviour worldwide. The realisation that life processes in the natural world are all indissolubly bound up with each other (so that, for example, Antarctic ice melt affects rainfall over the Amazon and convection currents in all the world oceans) is emphasised in David Attenborough's recent book *A Life on Our Planet* (London, Witness Books 2020). Change in any part of the environment will have effects on others, so we cannot regard animal, fish or plant extinctions as discrete events but as ones that set off a chain of reactions elsewhere, which cumulatively have massive climate, health and other consequences. This is what makes general environmental regeneration by all governments, as well as specific controls on emissions, such an urgent climatic as well as economic necessity at the present time.

Rewilding and environmental regeneration are the major priorities for diminishing the impact of climate change, rather than waiting for the technological solutions to be developed. These would be very welcome of course, but uncertainty about their effects and development time, plus their often-damaging side effects (mining lithium for electric car batteries for example or nuclear accidents) renders them risky options to prioritise as solutions. The world's worst nuclear accident at Chernobyl in Ukraine during the 1980s has a symbolic value here. The creation of a vast no-go area covering the polluted area has paradoxically allowed nature to return, with extensive reafforestation and a return of the original animals. The same process has occurred in the North East US, where small scale farms in hilly country have been unable to compete with mechanised, chemical-based agriculture in the Midwest and have been largely returned to natural woodland, with species previously on the brink of extinction now flourishing (wild turkeys for example) and no forest fires. Natural regeneration rather than simply planting trees without a supporting ecology diminishes climate change from the start. This makes it the most immediate and obvious strategy in tackling the problem – as well, of course, as halting its initial gestation.

8   Climate action in the world arena

8.1 Democratic action in the world arena

Even expert advice backed by a general scientific consensus and international religious fervour are likely to be severely constrained under an authoritarian regime focused above all on its own survival. Being independent of the regime can indeed be seen as blocking its pursuit of short-term advantages – mainly national economic growth and development – rather than expert advice being heeded (again an attitude shared by some democracies). If internal pressures are not strong enough to promote climate calming action, what about international ones? Here there are two problems where democratic action is key. The first is how to get democracies to adopt foreign policies that compel other countries to act in a constructive way. The second is how to put appropriate pressure on these in a world where 200 individual states have to coordinate their action if climate measures are to be effective.

Here we re-encounter the familiar problems of collective action already discussed in Chapter 2. Climate calming provides universal benefits from which no state or country can be excluded, even if it is a notable polluter and environmental despoiler. The world as a whole heats up or cools down regardless of national boundaries, so any one government can reason that it will gain anyway from general climate action if that succeeds, without foregoing any short-term benefits that come from emissions. This position has been most clearly taken up by democratically elected governments in Australia, the continent most sensitive to world climate change and already severely affected by it. They are opening new coal mines in the face of losing an eighth of their forest cover to a three-month fire. If a democracy in that situation can take the position it does, what are the prospects of getting more authoritarian regimes to abstain from environmental destruction?

DOI: 10.4324/9781003221630-8

Inside states, the problem of delivering public goods like regulation of antisocial activities has traditionally been solved by governments deciding what needs to be done in the public interest and enforcing compliance and payment by coercion if necessary. Were there a world government with similar powers, climate change could be tackled in this way, with emissions regulated and forests and oceans protected. Unfortunately, we only have approximations to world government at the present time, with no authoritative rule-enforcement and no central decision-making process. Even in the case of pandemics presenting an immediate threat, the World Health Organisation (WHO) can only issue general recommendations on how to react, with individual national governments deciding whether or not to modify them (or to follow them at all) inside their own territory. And this is with viruses that speed across national boundaries killing millions when they arrive! International economic and financial bodies like the World Trade Organisation (WTO) or World Bank have more clout because individual countries cannot break the rules without immediate consequences for their trade with other countries. But again, the rules themselves can only be set by getting a consensus among 200-odd members, which requires long drawn-out negotiation and compromises that hardly change anything. A recent example is the International Maritime Organisation (IMO) delaying any effective action on the use of highly polluting oil dregs as bunker fuel until 2030 – which will actually allow ship emissions to rise over the next decade! They already make shipping the sixth largest polluter on the planet.

This is where measures to defuse the world heat bomb require immediate action. But the 2020 World Conference that just might have negotiated them was postponed for a year. There is perhaps a glimmer of hope to be found in the international situation – the fact that the 200 states are not wholly independent but tied up together in various treaties and alliances for common action (though the IMO's lack of action is hardly encouraging). Some countries, however, are actively ceding powers to supra national bodies that make decisions on behalf of all members. The European Union is the prime example. Twenty-seven countries from the Atlantic to the Black Sea are delegating more and more powers to one central authority – a process driven by agreement between the two most powerful countries, France and Germany. And the European Commission, supported by the Parliament, is possibly the most advanced of all governments in terms of ecological and climate calming initiatives in the world today.

However, the Union itself has to get unanimous country agreement on its rules and one member, Poland, is one of the worst world offenders in terms of highly polluting emissions from its brown coal, and withdrawal of protections for its extensive forests. In Canada, a longstanding Federal Union, provincial governments have pressed ahead with a major ecological disaster, extracting oil from sands that require burning one barrel of oil to produce two. The constitutional division of power prevents Federal intervention here, even if the national government wanted to.

Canada as a whole and in its provinces, is of course a democracy, so internal protest and other pressures are possible. Under more autocratic regimes, however, protest is severely repressed. China is a case in point, viewing any querying of government decisions as political subversion.

8.2 The major international players

Beyond federal unions, another international reality is to be found in the existence of military alliances and strong economic relationships that tie most of the weaker, nominally independent states in the world to one or other of the superpowers that dominate its politics. Viewed from this perspective the 200-odd members of the United Nations reduce to around four – the European Union, the United States of America, Russia and China. The US and China can both exert economic and military pressures on other countries. The EU has massive economic clout but little military presence. Russia is the reverse, with massive military potential but a severely restricted economic base, which considerably limits its world influence. Getting four players to agree is, on the face of it, easier than getting unanimity among 200. Each has such strongly conflicting interests, however, that agreement is difficult. The EU and US have generally supported ecological action up to now, but both have patchy records. Russia has a considerable interest in exploiting its extensive natural resources, the major basis of its economy, and China is sucking them in for its manufactures. Russia, like Canada and Greenland, may even have an interest in promoting climate change, since melting permafrost opens its sub-Arctic regions to further exploitation.

The EU and US are full democracies, broadly speaking, and therefore have natural links with other democracies in the world, and Russia is (sort of) democratic. This leaves the way open for internal protests to push their policies in a climate calming direction, including foreign policies and international initiatives. Foreign pressure on China to do so as well is possible since the country has moved from outright hostility towards reluctant acceptance of incremental measures to avoid longterm climate change. Meanwhile, of course, it sucks in resources from neighbouring countries and is extending its demands for food and raw materials to Africa and Latin America, with massive deforestation the main consequence.

At the superpower level, China and the US are the main world players, and China is the main problem from a climate change perspective. The EU is weakened by its imperfect union. Russia can exert its full influence only regionally, with contained military interventions in southeast Europe and the Middle East. In its far eastern regions, it is mainly a supplier to the Chinese, and possibly threatened by extensive Chinese immigration into the Amur region.

8.3 China from an international and internal perspective

The Chinese Communist regime, while challenging the US for world supremacy, has two interlinked weaknesses. The first is internal. Its popular acceptance and hence internal stability are closely tied to its economic, social and military success over the last 40 years. Pure repression and surveillance in themselves may limit dissent, but they cannot contain it if general popular acceptance wears thin and the 'Mandate of Heaven' for the regime, based on its continued success, is withdrawn. One way it could be weakened is growing hostility by neighbouring countries (and even those further afield) at its penetration of their territory and assertion of its power (for example in neighbouring seas). There is an obvious precedent for such hostility in the widespread massacres of ethnic Chinese in Indonesia at the end of the 1950s. Vietnam is another example of a formerly dependent state increasingly resisting Chinese dominance once it won its war of independence in the mid-1970s.

Historically, almost all Chinese regimes have collapsed as a result of failures to contain or repel foreign invaders. None of its immediate neighbours are strong enough to actively threaten it – not even Russia or India. However, the United States, reacting to increased Chinese assertiveness, is increasingly involved in confrontations with the regime. Neighbouring tensions make it easier to organise a general international resistance, both economic and military, against it. The check – and even reversal – of economic growth that this would provoke might well encourage internal dissatisfaction and active dissent. One driver of this could be the increasingly severe pandemics that originate in China. These might ease population pressure, but they are hardly likely to promote popular confidence in the government.

Such immediate considerations could, if deployed effectively, induce the government to agree to – and actually adhere to – a rule-based international order that would extend to immediate climate calming measures. The EU has tried to broker this as a way out of the trade war between the US and China. Such rules would prohibit Chinese breaches of copyrights and patents as a means of promoting economic growth and abolish forced collaboration of foreign firms with Chinese firms as a way of gaining access to the internal market. These were tolerated when the Chinese world presence was less weighty, but they cannot be tolerated now that it is one of two or three major world players.

The same is true for restraint in environmentally damaging measures elsewhere, which might be one element in an economic peace agreement. China's increasing dominance of export markets is paradoxically also a basis for pressures on its government to change its ways. The more its general prosperity and internal standard of living depend on exporting, the more the country is susceptible to its customers (many of them democracies) closing their markets. This might hurt these democracies, but in the long-term is likely to hurt China more if its internal prosperity is severely affected. (About a third of Chinese industrial production is destined for export.) External pressure is especially potent when applied to an inherently fragile regime resting above all on economic success. Given the general need to stabilise the environment, this is an area where domestic agitation inside democracies to exert international pressure might influence global progress towards immediate climate calming developments.

There are positive advantages for China from beating its swords into ploughshares. Already a leader in solar power generation, it could penetrate world markets in this area in an acceptable and useful way. This would calm general suspicions about the dual functions of its technology for civilian use and also covert surveillance and communications and power disruption. These have already inhibited the export of electronic services to countries concerned about their own security. As such fears grow, they will block Chinese exports and growth anyway, so there is already a basis for agreement on tighter world trade rules and mutual inspection to ensure enforcement.

8.4 Towards a rule-based international environment

These world trade rules must include those for restructuring emissions and protecting the environment – but this can happen only if democracies insist on it. That is where domestic pressures from climate campaigners become relevant. It helps that democracies include most of the richest countries in the world among their members, with the largest consumer markets. In paying for imports they should set the rules, but only if they can be actively persuaded or pressurised into doing so, which is where appropriate domestic action comes in.

One avenue that democracies could take, but have surprisingly refrained from up to now, is to insist on all imports coming up to the same standards and regulations they apply to their own domestic enterprises. Following the logic of free markets (Figure 3.1), the rich consumer countries have gone for the cheapest goods however they have been produced – often by sweated labour in dangerous conditions. In part, this absence of controls has been fostered by an unwillingness to stand in the way of imports from poor countries that desperately need the revenues. But it has also been helped by their own efficient distribution chains, particularly with vast online suppliers like Amazon. These suppliers go for the cheapest goods on the world-wide free market so they can sell them a bit dearer, undercutting domestic products that are subject to labour laws, regulation, inspection and taxation. In turn this leads to over consumption and environmental waste as cheap clothing, for example, is worn once and thrown away, not to mention food.

A solution to many of these internal problems deriving from foreign under-regulation would be to insist on imported goods meeting better standards – possibly more relaxed than home ones and relative to local conditions, but still directed at raising quality of life and improving the quality of the product or service. For example, a minimum wage could be required for workers relative to the local standard of living. Controls on emissions and environmental degradation could be set. Enforcing these would require on the spot inspection by the importing country. Without these, goods would not be imported. A practical measure currently under consideration by the EU are border carbon tariffs. These are designed to make up the difference in price between imported goods, where this is due to laxer emission controls, and domestic goods made more expensive by tighter carbon restrictions.

These are actions in the international arena that democratic governments could take on their own without necessarily flouting WTO rules, all of which would exert significant pressure on foreign governments and do not require agreement by the Big Four. Another is to ban polluting ships from their ports regardless of what the IMO does not do.

The problem, as in other areas of necessary action, is getting democratic governments to act in the first place. However, action to 'create a level playing field' for domestic enterprise in the face of potential collapse after Covid-19 and with a moral justification from supporting workers' rights everywhere could create a powerful internal coalition with popular support for the initiative.

Of course, the same democratic pressures, as major consumers, can also be exerted on developing countries currently destroying their environment to provide a basis for economic growth. The stick and the carrot can both be employed here. Again, in their role as major consumers, the rich democratic countries can exert sanctions and support individual boycotts of, for example, South American beef, Indonesian palm oil, Australian coal and minerals, Canadian oil, wood from depleted forests and overfished species. Democracies are key here both as consumers and as open societies where domestic pressures for international action can be exerted. These are the sticks. The carrot is the minimum income guarantee already discussed, which could be offered in any country to producers and individuals who give up destructive practices. As suggested previously, this would have to be paid directly to individuals by an international administration operating on the ground to evade corruption and diversion of the funds by national agencies. This guarantee meets the demands of poorer countries for compensation from rich ones, while solving the problems of channelling monetary aid through inefficient and often corrupt local administrations.

National governments are bound to resist such a proposal. Most authoritarian regimes would see any organisation operating on their territory independently of their own government as a threat to their own standing and control. This perceived threat would be intensified in poorer countries where governments often use their control of international aid as a way to boost their own position (if not to line their own pockets). Even relatively democratic governments are likely to react adversely to direct and uncontrollable interventions from outside and present it as a threat to national independence – a manifestation of neo-colonialism on the part of ex-imperial powers. Picture Russia, for example, where even the humanitarian and truly international Red Cross is viewed with suspicion in official circles.

Supplementing opposition would be local landowners and political bosses, often linked to criminal organisations, who see their control threatened and their influence diminished by organisations coming in from outside and independently dispensing money. Since these are the very people profiting directly from environmental degradation, they are, of course, quite correct – independent direct individual payments would indeed remove the main levers of control from their hands and undermine their position, so they would fight it to the last under the banner of national (i.e. their) independence.

The prospect of independent direct income payments (if it is possible to publicise them properly within the country) would, on the other hand, have a galvanising effect on the poorer groupings in society, particularly in rural areas, but also in city slums and favelas. It would empower the indigenous groups who are often the last guardians of the forests in places like Amazonia and Indonesia. Above all, women with their own income would have increased autonomy, with immediate effects on birth rates and general demand for resources.

Utopian as it sounds, therefore, an internationally administered income guarantee, even if starting in only a few countries, is perhaps the most effective and immediate way of checking the reversal of climate calming processes throughout the world. Of course, the actual amount paid (as emphasised in previous chapters) would have to vary relative to the local cost of living. There is evidence that modest payments to the poorest individuals, within any society and across the world, raise life satisfaction sharply, but the effect diminishes as income goes up. The important thing is to have the minimum guarantee in the first place – details can be hammered out later on the basis of experience with pilot schemes.

The stick (or sword) has traditionally been the major implement used to shape international affairs to the wielders' satisfaction. This is still bringing on wars, famines and disease, particularly in the Middle East. Being highly localised, these can be tolerated outside the zone. The enormous destruction wrought by the two World Wars of the 20th century and the prospect of nuclear annihilation following on from them – precisely because of the dreadful worldwide threats they posed – have, on the other hand, strengthened progress towards a rule-based international system. Progress has been marked by the creation of inter-national institutions and agreements starting with the United Nations itself and its supporting organisations, the World Health Organisation, World Trade Organisation and so on.

With the World Climate Conferences (COPs) there has been an attempt to create similar regimes for the environmental and natural world – though, as we have seen, it has not been very successful. All previous international regimes rested on the positive benefits that they produced for their participants – primarily economic stability and avoidance of the depressions credited with being a prime cause of the Second World War. Economic growth and development being at their core accounts for environmental considerations being downgraded. But, at any rate, the carrots were there for most countries in sufficient quantity to overcome the collective choice problems that have come in at so many points in this discussion.

The major threat to the stability of the evolving rules-based system thereby established has always been military – the struggle for supremacy between superpowers and their allies, and in the second half of the last century primarily between the US and Russia. Thankfully the Cold War (1948–1985) never came to a head, given the prospects of assured mutual destruction it opened up, but it disrupted international rule-building, provoked destructive local conflicts, great environmental destruction (Indochina 1965–1974) and almost total neglect and indifference to earlier signs of climate change. For a brief spell (1990–2008) there seemed a prospect of surmounting the conflict and drawing the two major protagonists, the US and Russia, into a rule-based relationship, primarily on military, diplomatic and economic matters (but with some potential for the ecological side too). This was and is desperately important for climate change, given that Russia is the largest country in the world. Its relatively underdeveloped economy means that much of its area has survived as relatively untouched woodland and tundra – which, if preserved, will make a vital contribution to climate calming. However, this is now under threat from the general, and particularly Chinese, demand for minerals and wood. Russia's exploitation and devastation of its natural resources has been intensified by the economic sanctions imposed after the breakdown of its rapprochement with the West around 2010. The history of how this developed forms an object lesson in how wrong-headed, short-term considerations get in the way of dealing with the fundamental problems of environmental destruction, so they deserve some analysis here, particularly as their history may be repeating itself with the US and China.

8.5 Russia and the West: from confrontation to cooperation – and back again?

In the 1950s the old Russian Empire, under the new name of the Soviet Union and dominating the whole of Eastern Europe, began to emerge from a brutal period in which its single party communist regime had maintained its rule through repression and terror. These were slowly relaxed from the mid-1950s to the early 1980s. This was also the period of the Cold War in which a western military and diplomatic alliance (NATO – the North Atlantic Treaty Organisation), led by the US, confronted the Soviet Union and the countries it dominated. Both alliances had nuclear weapons, massive military resources and bases throughout the world. They never came to full scale combat, but fought each other covertly in a series of wars and local confrontations. The US managed to draw ahead after detaching China from support of the Soviet Union in the 1970s and expanding both economically and in nuclear-related technology during the same period. The government that came to power in the Soviet Union in the early 1980s decided to end the conflict that was sapping all their resources. It negotiated a series of treaties with the US that resolved the main issues between them in the mid-1980s.

In the face of internal tensions, the Soviet State then embarked on an extraordinary relaxation of foreign and internal controls, first withdrawing support from subordinate governments in Eastern and Central Europe, then by giving independence to most of the non-Russian republics of the Soviet Union itself – ceding about a fifth of its old territory. This was an unprecedented act of voluntary withdrawal, equivalent to the US giving back to Mexico its southern border states from Texas to California, or China freeing Tibet, Sinkiang and Mongolia rather than colonising them.

The western alliance, however, never acknowledged the magnitude of this gesture, partly because Russian withdrawal was followed by ten years of political confusion and economic collapse within Russia itself that helped to disqualify it as a credible negotiating partner. The Russian regime did, however, reconstitute itself as a democracy during this period with regular elections, autonomous parties competing in them and a relatively free press.

When a more coherent government was installed at the beginning of the new century, it maintained democratic constitutional practices and pursued more stable policy lines. The western alliance recognised this by admitting Russia to the meetings of the Group of Seven, the leading economies of the world, which then became the Group of Eight in spite of Russia's economic weaknesses. This gesture compensated to some extent for NATO – a military alliance against Russia – signing up ex-Soviet Republics as members, in breach of tacit agreements earlier. The West went on to intervene directly in elections in Ukraine, Russia's nearest neighbour, and in Georgia in the Caucasus, and gave further indications of recruiting them into western alliances. Russia reacted by active military interventions along its border that provoked western economic sanctions. The Group of Eight returned to being the Group of Seven. Important foreign policy interventions in Libya and Syria were taken without reference to Russia, which went on to oppose western military initiatives, though in a contained way, in the Eastern Mediterranean and the Black Sea. Meanwhile, cyber warfare opened up with the US, supplemented by clandestine operations and political assassinations on both sides – widely perceived as opening up a new cold war.

That relations have come to this seems primarily due to western mishandling of a situation in which Russia had made immense political and military concessions for which it got little or no reward. Russia's main concern after 2000 was to be consulted over world decision-making after being excluded from it in the 1990s. Recognising that this had been due to its own internal weaknesses, it tried to remedy these and play along as a partner in world decision-making forums in the early 2000s. However, its exclusion from crucial decisions and continued western expansions of influence and alliances into its 'Near Abroad' provoked it into hostile – but very localised and contained – reactions from 2010 onwards. These were blown up in the West into precursors to a new cold war that has taken up increasing policy attention and resources over the last decade.

8.6 Lessons for the future

This sorry history is of major relevance to the environmental and climatic situation because:

a   It becomes a major short-term issue diverting attention from the cataclysmic but longer-term problem of climate change;

b   This is a particular problem with Russia. Under pressure from western sanctions, deforestation has speeded up. Processes of change, such as penetration and pollution of the Arctic Ocean, may actually be welcomed internally as giving access to more natural resources to exploit;

c   The West's negative reaction to Russia's massive opening-up and limited adoption of democratic procedures gives little encouragement to countries like China if they want to do the same. Indeed, China's own progression from a more open to a more closed and repressive political system over the last decade could well result from what they have seen happening between the West and Russia. That has already sparked off the possibility of a new Cold War with China on a much wider front, as opposed to Russia's contained regional responses This is now taking up attention from governments and policy-makers and diverting them from climate change;

d   Acceptance of a rule-based international system depends heavily on the most powerful players seeing benefits from it. Sanctions may be applied to those who infringe the rules, but they in turn do depend on all the other countries agreeing and supporting them. Where one rule is applied to, for example, territorial annexations by Israel or Turkey and quite another to Russia in Ukraine, the rules themselves cannot be regarded as impartial. But we desperately need impartial, binding, international rules related to climate calming if we are to achieve it

This overview of international relationships over the last 70 years gives a clear indication of what climate activists inside the established democracies should be aiming at when it comes to their foreign policies. First comes an end to the cold war developing between NATO and Russia – which is at one and the same time a massive distraction from the fundamental problem of climate change as well as a massive diversion of the resources that should go to tackling it. The same can be said of the developing confrontation with China. At least this is at an earlier stage, which renders it possibly easier to resolve. More clearly than Russia, China is itself highly vulnerable to the effects of climate change and should therefore be more motivated to perceive benefits from joining in and tackling it.

Besides protesting and withholding tax support for growing rearmament programmes, democratic climate campaigners could also work towards getting a universal minimum income guarantee both at home and abroad. Particularly in the developing countries, this should remove the need for the poorest to destroy their own environment in order simply to carry on. It could also undermine the position of leaders and parties urging economic development at all costs, and instead turn their energies into a sustainable agriculture complementing instead of destroying its environment. Costa Rica is a shining example where the rainforest has been restored to become a focus of enlightened tourism and external income. Climate campaigning should be directed towards replicating Costa Rica's achievement everywhere, using the domestic and international strategies outlined in this book.

8.7 Further reading and reflection

The Consumption of a Finite Planet, already referred to in the discussion of a minimum income guarantee in Chapter 3, points to the impossibility of bringing the whole world up to the Western standard of living and the need to adapt both the level of a minimum income guarantee and its direct international administration to the realities of local situations. Ian Budge and Sarah Birch provide an in-depth description of how to create a level playing field for domestic industry by extending worker and environmental standards to imported goods in Budge and Birch (2014), *National Policy in a Global Economy* (Palgrave MacMillan, Basingstoke UK).

The fragility of the Chinese regime in the face of economic slowdown is underlined by a recent report in August 2020 that national unemployment is actually 20.5 percent of the workforce (70–80 million) rather than the official 6 percent, with little or no social support for those affected (*The Week*, 8 August 2020 p 18). This is described as a ticking time bomb that 'threatens to break the social contract based on unfaltering political obedience in exchange for economic enrichment'. Clearly Western democracies should seek to strengthen the internal experts pointing this out to the Communist leadership, rather than initiating another cold war, thus reinforcing a rules-based international environment that faces up to the string of natural disasters now threatening us.

Epilogue

A recent article headed 'Why We Fail to Prepare for Disasters' (*The Week*, 1 August 2020, p 48) provides a chilling qualification to the initiatives urged here. It lists a series of events, from the flooding of New Orleans in 2005 prefigured by Hurricane Ivan (2004) to the series of pandemics from Sars (2003) to Mers (2015) preceding Covid-19. All gave ample warning of more to come, but these were simply ignored by politicians and public.

Psychologists describe this behaviour as normalcy bias. We have survived preceding disasters so we will probably survive the next. People are confused, do not know what to do in the face of impending danger and so carry on as normal. As noted at many points in this discussion, politicians want to deal with short-term issues rather than long-term dangers (these are for the next government). The continuing and ever extending process of climate change may have more of an impact in warning us of dangers to come. History tells us maybe not. We can but hope and act now by giving both public and politicians clear alternative courses of action to follow, as laid out above.

All events, however, have a positive as well as a negative side. The positive side of Covid-19 is that it has had extended effects not only on social life through lockdowns and quarantines, but also disruptive effects on the existing economy and employment, spilling over to politics as well. It is a major wake up call, more so than anything coming before. It also has the potential to undermine existing ideologies and thinking, in ways which strengthen campaigns for climate action.

First of all, the need to avert massive hardship and unrest has produced a series of measures to make up incomes. These set precedents for the permanent minimum income guarantee discussed at many points above. Clearly necessary just to keep things going (including free markets), the action undermines all the neo-liberal arguments set out in Box 3.1 by making clear that free markets and private economic

DOI: 10.4324/9781003221630-9

growth are only products of government actions and guarantees. In the current situation, control and regulation need to be extended rather than restricted.

Unprecedented government borrowing to finance its emergency measures also destroys concerns about 'balancing the books' by cutting public services or abandoning regulations and environmental action. Since these provide jobs and thus are central to getting the economy going again, an environmental 'New Deal' along the lines of 1930s America is, on the contrary, the only way to sustain financial confidence. When confidence is strong governments have unlimited credit so they can undertake all the environmental initiatives necessary. Environmental restoration also provides a clear plan, otherwise lacking, to boost general confidence, particularly when supported by a minimum income guarantee. In combination, these initiatives are also necessary to maintain jobs and with them political stability and security.

To ensure that governments do act appropriately, climate campaigners have a powerful lever in that they can now threaten the financial confidence that governments have to maintain to simply go on functioning. They can do so both by preparing and publicising entirely legal tax avoidance schemes (only to be effected if governments go down the wrong road) and by winning over established parties' voters. Increasing vote volatility in an ever changing and threatening world gives climate campaigners the chance to do this. If we just sit back and let events happen, they will happen: but if we organise and innovate now, we can change them.

The bright side to all this is set out in David Attenborough's inspiring 'witness statement' (*A Life on Our Planet*, 2020 Witness Books, London). Mostly, we only have to leave off attacking the environment in order to let its self-adjusting interactive processes restore themselves as they have after recent disasters like the nuclear meltdown at Chernobyl. Going back to an older 'normal' than the one just before Covid-19 is not so hard to achieve, but we must achieve it now, restructuring our economic and political life along the lines we have indicated in order to carry it through in time.

Bibliography

Both our knowledge of and political responses to the climate crisis have moved on so rapidly over the last decade that they have left most earlier research or writing behind. Consequently, it seems best, in a short and tightly focused book like this one, to concentrate on more recent writing over the last ten years. Anyone wishing to go further back in the debate will find copious references to earlier literature in the books cited here.

Attenborough, D. (2020) *A Life on Our Planet* (London: Witness Books).

Budge, I. and Birch, S. (2014) *National Policy in a Global Economy: How Government can Improve Living Standards and Balance the Books* (London: Palgrave Macmillan).

Budge, I. (2019) *Politics: A Unified Introduction to How Democracy Works* (London: Routledge).

Bulkeley, H., Andonova, L., Betsill, M., Compagnon, D., Hale, T., Hoffmann, M., Newell, P., Paterson, M., Roger, C. and VanDeveer, S.D. (2014) *Transnational Climate Change Governance* (Cambridge: Cambridge University Press).

Compston, H. (2012) *Climate Clever: How Governments Can Tackle Climate Change (and still win elections)* (London: Routledge).

Dessler, A and Parson, E.A. (2009) *The Science and Politics of Global Climate Change: A Guide to the Debate* (Cambridge: Cambridge University Press).

Dorling, D. (2020) *Slowdown: The End of The Great Acceleration - and why It's Good for the Planet, the Economy, and Our Lives* (New Haven, Conn: Yale University Press).

Funk, C. (2021) *Drought, Flood, Fire: How Climate Change Contributes to Catastrophes* (Cambridge: Cambridge University Press).

Giddens, A. (2009) *The Politics of Climate Change* (Cambridge: Polity Press).

Gough, I. (2017) *Heat, Greed and Human Need: Climate Change, Capitalism and Sustainable Wellbeing* (Cheltenham: Edward Elgar Publishing Limited).

Harris, P.G. (2021) *Pathologies of Climate Governance: International Relations, National Politics and Human Nature* (Cambridge: Cambridge University Press).

Hoffman, M. (2011) *Climate Governance at the Crossroads: Experimenting with a Global Response after Kyoto* (Oxford: Oxford University Press).

Klein, N. (2014) *This Changes Everything: Capitalism vs. The Climate* (New York: Simon & Schuster).

Kolbert, E. (2021) *Under a White Sky: The Nature of the Future* (London: Penguin Random House).

Malm, A. (2020) *Corona, Climate, Chronic Emergency: War Communism in the Twenty-First Century* (London: Verso Books).

McKibben, B. (2019) *Falter: Has the Human Game Begun to Play Itself Out?* (London: Wildfire).

Monbiot, G. (2016) *How Did We Get into This Mess?: Politics, Equality, Nature* (London: Verso Books).

Monbiot, G. (2017) *Out of the Wreckage: A New Politics for an Age of Crisis* (London: Verso Books)

Newell, P and Paterson, M. (2010) *Climate Capitalism: Global Warming and the Transformation of the Global Economy* (Cambridge: Cambridge University Press).

Piketty, T. (2019) *Capital and Ideology* (Cambridge, MA: Harvard University Press).

Pretty, J. (2013) *The Consumption of a Finite Planet: Well-Being, Convergence, Divergence and the Nascent Green Economy* (Dordrecht: Springer Nature).

Raworth, K. (2017) *Doughnut Economics: Seven Ways to Think Like a 21st Century Economist* (New York: Random House).

Sikka, P. (2013) Smoke and mirrors: Corporate social responsibility and tax avoidance. *Accounting Forum*. 37(1), 15–25.

Sikka, P. (1997) *Accounting for human rights: The challenge of globalization and foreign investment agreements*. Critical Perspectives on Accounting. 8, 149–165.

Victor, D. (2011) *Global Warming Gridlock: Creating More Effective Strategies for Protecting the Planet* (Cambridge: Cambridge University Press).

Vollrath, D. (2019) *Fully Grown: Why a Stagnant Economy is a Sign of Success* (Chicago: University of Chicago Press).

Wallace-Wells, D. (2019) *The Uninhabitable Earth: Life After Warming* (London: Penguin Random House).

Willis, R. (2020) *Too Hot to Handle?: The Democratic Challenge of Climate Change* (Bristol: Bristol University Press).

Newspapers, magazines and journals

The Economist www.economist.com

Essex County Standard essexcountystandard.co.uk

The Guardian www.theguardian.com

The Herald www.heraldscotland.com

The Herald on Sunday www.heraldscotland.com

The Independent www.independent.co.uk

London Review of Books www.lrb.co.uk
Mother Earth www.soilassociation.org/motherearth
The Scotsman www.scotsman.com
The Week www.theweek.co.uk

Official sources (UK)

Climate Change Committee Reports (www.theccc.org.uk)
Progress Report to Parliament (June 2021)
Independent Assessment of UK Climate Risk (June 2021)
Sixth Carbon Budget (December 2020)

Relevant websites

www.investopedia.com
https://www.sunrisemovement.org/home/
www.350.org
https://www.climaterealityproject.org
https://extinctionrebellion.uk/
https://www.ran.org/
www.greenparty.org.uk
www.ienearth.org/
www.greenpeace.org.uk
https://takeclimateaction.uk/
www.movimento5stelle.it
https://climatenetwork.org/
https://greens.scot/
https://scientistrebellion.com/
https://europeangreens.eu
www.theclimatecoalition.org
www.campaigncc.org
legislation.gov.uk
https://green-alliance.org.uk/
https://naee.org.uk/
https://energysavingtrust.org.uk/
https://friendsoftheearth.uk/
www.ethicalconsumer.org/
www.wildlifetrusts.org/
www.recyclenow.com/
www.renewableuk.com/
www.the-ies.org/
www.cpre.org.uk/
https://elflaw.org/
www.climateneutralgroup.com
www.climateassembly.uk/
www.wwf.org.uk

Index